Understanding House Construction

Second Edition

John A. Kilpatrick

HOME BUILDER PRESS

Home Builder Press®
National Association of Home Builders
1201 15th Street, NW
Washington, DC 20005-2800

Understanding House Construction, second edition
ISBN 0-86718-382-9

© 1993 by Home Builder Press® of the National Association of Home
Builders of the United States of America

Cover photo © 1992, Tom Campbell, FPG International Corp.

Printed in the United States of America

Library of Congress Cataloging-in-Publication Data

Kilpatrick, John A., 1954-
    Understanding house construction / John A. Kilpatrick. — 2nd ed.
        p.   cm.
    Rev. ed. of: Understanding house construction / National
Association of Home Builders.
    Includes bibliographical references.
    ISBN 0-86718-382-9
    1. House construction. I. National Association of Home Builders
(U.S.) II. Understanding house construction. III. Title.
TH4811.K49   1993
690'.837—dc20   92-35117
                                                            CIP

For further information, please contact—

Home Builder Press®
National Association of Home Builders
1201 15th Street, NW
Washington, DC 20005-2800
(800) 223-2665

1/93   HBP/AUTOMATED   6M

Contents

Illustrations

About the Author

John A. Kilpatrick is a writer on homebuilding and real estate topics from Columbia, South Carolina. He holds an M.B.A. and is currently research associate in the Office of Research at the University of South Carolina, where he is also a candidate for a Ph.D. in Real Estate Finance. He was previously president and broker-in-charge of a development and construction firm and has served in a variety of positions in the homebuilding industry.

Acknowledgments

Special thanks go to Dr. Jerry Householder, Chairman, Department of Construction, Louisiana State University; Thomas Kenney, Director of Contract Research, NAHB Research Center; Donald Luebs, Director, Building Technology, NAHB Research Center; Richard Morris, Senior Technical Adviser, NAHB Technology and Codes Department; and Dr. Leon Rogers, Associate Professor of Technology Education and Construction Management, Brigham Young University, for their thoughtful and thorough review of the manuscript. In addition, thanks go to Richard Meyer, Director, NAHB Technology and Codes Department, for his thoughtful review of the outline for the book.

The author also wishes to thank the staffs and management of Tidewater Plantation in North Myrtle Beach, South Carolina; Prudential - MBF Realty in Myrtle Beach, South Carolina; Rymarc Homes by Marc Homebuilders, Inc., of Columbia, South Carolina; and Rose Anne O'Reilly of the Horry-Georgetown, South Carolina, Home Builders Association for their valuable assistance on this book.

This book was produced under the general direction of Kent Colton, NAHB Executive Vice President, in association with NAHB staff members James E. Johnson, Jr., Staff Vice President, Operations and Information Services; Adrienne

Ash, Assistant Staff Vice President, Publishing Services; Rosanne O'Connor, Director of Publications and Project Editor; Melissa Brown, Publications Editor; Carolyn Poindexter, Editorial Assistant; and David Rhodes, Art Director.

Disclaimer

Introduction

Almost nothing in the world is done the way it used to be done. However, if you think about it, many things these days are actually done better than before.

Take the typical new home. It's a far cry from the one your grandparents grew up in. Try finding enough receptacles in a 50-year-old kitchen to plug in a modern refrigerator, microwave oven, coffeemaker, television set for the morning news, and the toaster oven for that rushed morning breakfast. Then you better have some extra fuses around for the antique fuse box if you plug in all these appliances at the same time!

Today's new homes are designed to balance amenities with maximum affordability, and they contain features and fixtures planned for the convenience of the modern, often two-wage-earner family. Most are better insulated than ever before and, particularly in the southern states, have air conditioning.

The labor of dozens of skilled artisans, tradespeople, and professionals contribute to the construction of a new home. From the engineers to the designers or architects who may shape or create the house plans, from the workers who clear the site and lay the foundation to the framers and roofers, from the plumbers and electricians to the painters and landscapers, all have important roles to play in constructing a home.

A new home also requires materials produced by many specialized suppliers and many behind-the-scenes activities like construction lending. However, the person who can make it all come together through careful coordination is the experienced professional home builder.

Fifty years ago, skilled artisans constructed a home from raw lumber, shipped to the job and assembled on site. If the weather was rainy or cold or the suppliers couldn't get materials to the site that day, then the skilled artisans went somewhere else. Today, many components of a new home are constructed off site with the latest manufacturing techniques and technology.

Weather, utility delays, and even nighttime darkness are now much less apt to slow down the construction process. There is little or no wastage. The materials stay dry and protected in shops or warehouses until needed, and not much detail work is done on the job site. The components of the home are packed on a truck, shipped to the site, and assembled in the shortest possible time. Onsite portable electric generators are widely used to provide power for such construction tools as saws, air compressors, and nail guns.

Photo I.1 shows a flatbed trailer that contains all of the panelized components needed to frame the walls of a three-bedroom, two-bath home. The roof truss components lying next to the trailer complete the framing package. Photo I.2 shows a home that a carpentry crew has just roughed in from materials shipped on a flatbed trailer.

Modular and Panelized Construction

Component- or systems-built housing uses various forms of factory-produced items that range from simple components like roof trusses to completely prebuilt modular units. The two major types of systems-built housing are modular and panelized. Log and dome homes are also subcategories of systems-built housing.

Modular homes consist of one or more three-dimensional modules or preassembled boxes that are about 90 percent complete when the manufacturer ships them from the factory. Once the modules arrive on site, builders place two or more sections on the foundation to create the finished home. Builders can even stack the modules to make two- and three-story homes.

I.1 Delivery of Framing Components to a House Site

I.2 Recently Constructed Component Home

Panelized components look like the unassembled parts of a box and consist of panels about 8 feet high and from 4 to 40 feet long. Roof and floor trusses, prehung doors and windows, siding, and interior trim are often sold with wall panels as complete house packages. Log home manufacturers mill and precut logs for specific house plans, and dome home manufacturers make panels that builders can assemble into domes.

Although modular and panelized construction is gaining popularity, this book focuses on stick-built construction techniques because they include all of the steps required in building a house. As this book explains house construction, it also defines some of the terms in the process and allows you to peek behind the scenes into the way things are done. However, this book is *not* intended to substitute for the services of an experienced professional home builder. It is also not intended as a short course on building your own home or as a guide to the specifics of the construction of your home. Out of necessity, building conditions, materials, techniques, and codes vary among regions of the country.

In fact, because of varying regional conditions, building codes regulate construction at the local level in most areas. These code requirements depend on such local factors as climate, wind, geological features, soil conditions, and even the availability of certain resources. As one example, certain building features, such as basements, that are common in some parts of the country are impractical to build in others.

Building Codes

Generally, local building codes govern building, plumbing, heating and air conditioning, electrical systems, and fire safety. While a few municipalities (mostly major cities) write their own codes, most state, county, and local jurisdictions adopt model codes prepared by one of three major model code service organizations:

- Building Officials and Code Administrators International (BOCA)
- International Conference of Building Officials (ICBO)
- Southern Building Code Congress International (SBCCI)

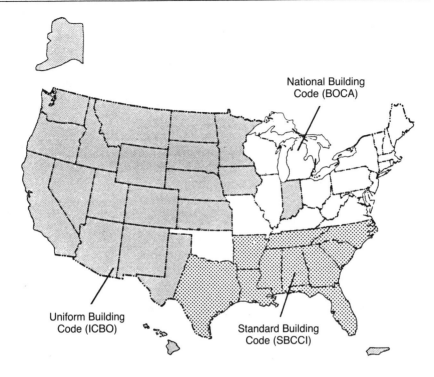

National Building
Code (BOCA)

Uniform Building
Code (ICBO)

Standard Building
Code (SBCCI)

I.3 Regions of Model Code Influence

BOCA's code is called the National Building Code; ICBO's, the Uniform Building Code; and SBCCI's, the Standard Building Code. Each group's codes influence a different region of the country: BOCA in the Northeast and Midwest; ICBO in the West; and SBCCI in the Southeast. (See drawing I.3.)

The Council of American Building Officials (CABO) is a federation of BOCA, ICBO, and SBCCI, which publishes the One- and Two-Family Dwelling Code. The CABO code influences the entire United States because it is recognized by BOCA, ICBO, and SBCCI. Building officials and others with firsthand knowledge of construction practices develop all these codes. However, for building codes to become law, they must be adopted by a local legislative body.

Health codes, which are established and maintained at the county or municipal level in most parts of the country, govern wells and septic systems. County or municipal building or engineering departments usually control public water and sewer. In addition, for electrical work, all localities must follow the National Electrical Code (NEC),® established and administered by the National Fire Protection Association.

At the beginning of the process, the builder obtains permits from various authorities to proceed with construction. As the house is built, the local building and health departments inspect the process to ensure that the house meets code requirements. The primary concern of the local authorities is to assure the health and safety of the public—in this case, the future occupants of the home. You should be aware, however, that inspections do not evaluate the quality of construction—only compliance with applicable codes.

Inspections

After building permits have been issued and construction begins, inspections are required at specified stages of completion. The builder informs the appropriate municipal department when the house is ready for inspection. The inspector then conducts the inspection and any failed items must be corrected and reinspected.

While inspection requirements vary from community to community, the following inspections are typical:

- The local municipal sewer and water authority or engineering department inspects the municipal water service and sewer connections, unless the builder installs a well and septic tank system, in which case the health department conducts the inspection.

- The building department inspects footings, open trenches, and formwork before the concrete is poured. If the builder uses steel reinforcement, it is inspected at the same time. Footing depth and soil conditions are checked to ensure that the footings can provide adequate support for the house.

- The building department then inspects the completed foundation before waterproofing and backfilling.

- The building department inspects roughed-in framing, plumbing, electrical wiring, heating and air conditioning ducts, insulation, and other items before the builder closes in the walls. Often specialized inspectors inspect building, plumbing, and electrical systems separately.

- The building department performs a final inspection to check plumbing, electrical, and mechanical systems; interior and exterior finish; and landscaping. If everything is in order, a certificate of occupancy is issued.

As you can see, the process of obtaining permits and meeting code requirements is as complicated as building the house itself. In fact, an experienced professional home builder is needed to sort through the maze of techniques, requirements, and specifications for taking a house from prints to closing. As the book describes the steps involved in building a home, it can become your guide to the construction process and translate the language of homebuilding into something you can easily understand.

1

Site Preparation

Choosing the site is the very first step in building a house. It influences not only building techniques but almost every other aspect of the homebuilding process.

Selecting the Site

Slope, access, drainage, vegetation, and many other site characteristics can affect the style of a home. A highly sloped lot may favor a split-level home. A broad, flat lot may favor a ranch-style home. Nearby water or underground rock may prohibit construction of a basement, while a narrow access may prevent the delivery of large components such as beams or broad glass panels.

Since a home is designed with the site in mind, once the site is selected the builder must consider the following:

- location of the home that makes best use of the natural grade and slope, keeping in mind drainage and aesthetics
- natural features, such as foliage, bodies of water, and other site geology, that enhance the final landscaping
- position of the home on the site relative to other features to reduce construction costs
- direction of the home that makes best use of the location of the sun, balancing energy conservation with aesthetics

If all else is equal, a home on a perfectly flat lot will cost less to build than a home on a highly sloped lot. This is because the costs are higher for the bricks, mortar, concrete, and labor required to build the larger, irregular foundation for a sloped lot. With some sloped sites, the builder may need to rent a crane or other machinery to lift materials that would have otherwise been moved by hand.

Determining Slope and Drainage

More than any other factors, the slope and drainage of a site affect the construction of the foundation. For example, a flat site is most practical for a slab-on-grade home discussed in Chapter 2. In contrast, a steeply sloped lot lends itself to a home with a basement because the depth of a basement allows the grade to angle down the sides of the house and assist with basement drainage. (For examples of basement drainage and slopes for basement drainage, see photo 1.1 and drawing 1.2.)

1.1 Basement Drainage

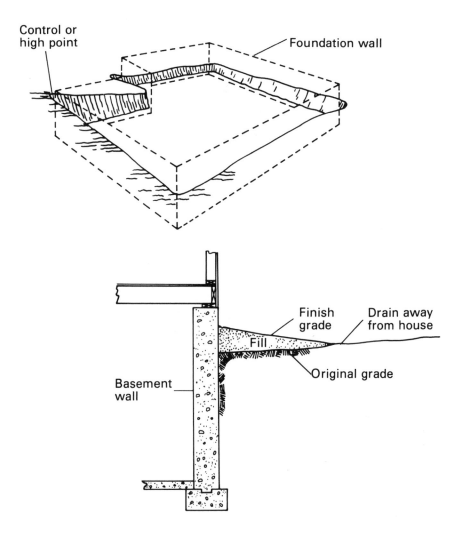

1.2 Slopes for Basement Drainage

A flat site or a site with a gentle slope permits the natural drainage needed for the construction of a home with a crawl space. A crawl space is an unfinished area between the ground and first floor large enough for a person to crawl through. A crawl-space home often uses a pier-and-curtain-wall foundation. This type of home sits on piers that are then screened by a low wall around the perimeter of the foundation.

Other site features that affect the type and construction of the foundation include—

- rock outcroppings
- shallow depth to bedrock
- shallow groundwater
- location of natural drainageways through the site
- dense vegetation

From the beginning the builder needs to plan the final landscaping and erosion control. Regardless of the size, cost, or style of the home, these are serious considerations. Whether a new homeowner has a $10,000 landscaping budget or plans to sow a handful of grass seeds, the final grading of the site can affect the appearance and long-term value of the home.

Addressing Legal Considerations

Virtually all areas of the country have agencies such as zoning boards, planning agencies, planning councils, or development boards. Many subdivisions also have architectural review boards provided for in their charters and mandated in the deed restrictions for the lot. Federal, state, and local environmental restrictions may also apply.

What does all this mean for the builder? Plainly put, local officials want to ensure that all homes and businesses fit into a cohesive master plan, that shared resources such as utilities and storm drainage are fairly allocated among landowners and dwellers, and that community neighborhoods are protected from inappropriate land use. In many cases, this may require careful integration of new homes into existing neighborhoods. (See drawing 1.3.)

A lot usually has utility easements running along the front and possibly the side and back edges of the yard. A utility easement is the right of the local utility companies to run water, electric, gas, telephone, and sewage lines through certain areas of the property.

The local utility companies have more or less free access to install, maintain, and replace utility lines either above or under the ground in easement areas, so a builder should not

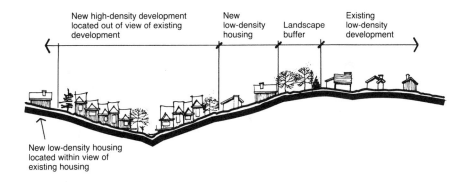

1.3 Cross Section of Proposed and Existing Development

build anything permanent within these areas. For example, if a homeowner plants prized gardenias in an easement area, he or she may come home one day to find a utility truck parked in the middle of them.

Utility easements can even cross the middle of some lots. As a result, when developers plan a subdivision, they realize that some lots are less desirable for building due to the need to place underground water, sewer, or drainage lines through them.

A lot can be less desirable for building for other reasons as well. Dune lines or mean high-tide setback lines, which prohibit building on certain low-level lots, affect many coastal neighborhoods. Some lots in such areas may even be lower than the 100-year floodplain designated by the federal government, which may render a home on such a lot uninsurable. A floodplain is land that may become covered with water when floods occur and the 100-year floodplain is the height to which water is anticipated to rise within a 100-year storm event.

None of these site conditions is necessarily obvious. The builder must do a bit of research and carefully examine a development site to determine its appropriate use. Once the builder has formed an opinion that a site is buildable and determined the usage or architectural restrictions that may apply to the site, the builder must then navigate through the permitting process. This process is often complex and can become time-consuming and costly. Although the permitting process can take from one month to a year or more, an experienced builder can steer the plans through the process and obtain permits in the shortest time possible.

As part of a permitting process, a zoning or planning board must certify that a proposed home is within the bounds of the local zoning ordinances. An architectural review board in a subdivision may have to attest to the aesthetic quality of the design for the home. Members of this board may want to look at such things as plans, elevations, paint colors, designs for landscaping, and samples of siding or roofing materials. The utility board may want to certify that water and sewage service is available.

A building official will have to certify that the house plans meet all of the local building codes. If the builder or homeowners want to drill a well or install a septic tank on the site, they will have to receive advance permits from the health department. The cable television company may even become involved in issuing a permit for the new home.

Preparing the Site

Once the legal hurdles are crossed, the builder begins the physical process of constructing the home. The first step is clearing the actual house site and driveway for the delivery trucks. At this time, the builder may also remove large trees and other vegetation as needed, perform rough grading for the eventual landscaping, and make early provisions for anticipated drainage patterns.

The builder will then stake out the house by locating and marking all of its corners and ensuring that its final position matches the site plans. (See photo 1.4.) Stakes are then driven into the ground to mark the exact location of the house. At this point the builder also needs to take care of several housekeeping chores, such as locating the temporary power source, marking the location of eventual water and sewer taps, and working with the public utility company to ensure that no additional easements are necessary.

1.4 Stake-out

So, after selecting a site, the builder can plan the house on the site to achieve the best use of the natural grade and slope for drainage and aesthetic purposes, while at the same time keeping construction costs down. The builder must then complete the permitting process before preparing the site for construction. Once through the permitting process, the builder can clear and grade the site and stake out the house. Then the next major step in building a house is constructing the foundation.

2

The Foundation

Home builders have an old saying that, if you start out with a level, square foundation, then almost everything after that is cosmetic. Chapter 1 discussed planning and preparing the site for the foundation. Now Chapter 2 will explain the actual design and construction of the foundation. Although the discussion will focus on masonry foundations, it will also touch on pressure-treated wood foundations. The first step in building the foundation is constructing the footings.

Constructing the Footings

The builder begins by excavating for the foundation and marking the location of the footing trenches. (See drawing 2.1.) The footings form a continuous concrete pad on which the foundation walls rest. The builder usually digs the footing trenches around the perimeter of the house and extends them several inches or more on either side of the location of the future foundation walls. If the site has a slope, then the builder must construct stepped footings to follow the grade of the slope. Under some conditions, the builder may use lumber forms to construct the footings on the ground. After the concrete sets, the builder then removes the forms.

The builder may construct a rough frame of batterboards at the corners for the house. (See drawing 2.1.) The builder then marks the location for the exterior foundation walls

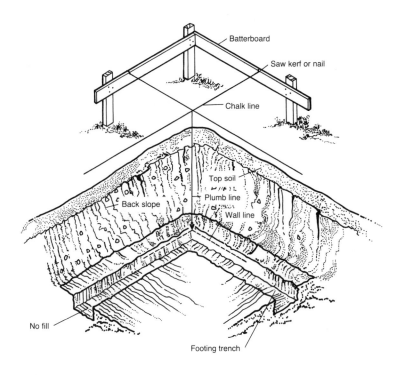

Batterboard

Saw kerf or nail

Chalk line

Top soil

Plumb line

Back slope

Wall line

No fill

Footing trench

2.1 Excavation of House Corners and Footing Trenches

with heavy twine between the batterboards. Using this temporary life-size pattern, the builder makes sure that the
house layout is square and the cleared footprint of the house
is level.

Brick and stone masons will use the lines marking the
shape of the house to build the foundation. If the foundation
is even slightly out-of-square, then everything above it will
also be out-of-square. This is no small matter, so the builder
takes great care to ensure the foundation is square. Only
after the builder and often a local building inspector are satis-
fied with this layout can the workers begin excavating the
foundation and digging the trenches or building the forms
for the footings. (For a picture of footings under construction,
see photo 2.2.)

The builder has other trenches dug at this time as well. If
the house has a crawl space or a basement foundation, then
the interior pillars or columns require footings. For these the

2.2 Construction of Footings

builder digs square or rectangular holes, somewhat larger than the proposed pillars. The builder, along with the designer or engineer, determines the proper size and depth of all trenches based on several factors, including local soil conditions and building codes. Again, a local inspector may be required to certify that the trenches or forms are properly prepared to receive the concrete footings.

If a home has a slab-on-grade foundation, which is a layer of concrete poured directly on prepared soil, then the builder digs the trenches for the plumbing and electrical lines at this time. Before the slab is poured, the builder calls in the plumber and electrician to install the rough plumbing lines and the conduit for the electrical lines. Some slabs are poured at the same time as the footings and some are poured separately. A variety of climate and site conditions, building techniques, and local codes can influence when footings are poured.

Local codes and soil conditions may require the builder to reinforce the footings with steel bars, often called reinforcing bars or rebars. These bars are usually laid horizontally, running the entire perimeter of the footing. The builder often raises these bars up a few inches off the bottom of the footing trench and rests them on metal or masonry supports. In areas of high wind or seismic stress, local codes may also require that a vertical rebar connect the footing with the foundation. Otherwise, the foundation or slab simply rests on the footing.

Once the concrete footing is in place, it needs to cure or harden for a short period before construction can continue. Usually a day or two is sufficient for the concrete to cure, but the curing process may take longer in colder weather. In fact, in very cold weather, the builder needs to take special precautions to keep the ground and concrete from freezing.

After the footings have cured, the builder can then move on to the next steps in constructing the foundation. The type, materials, and design of the foundation will affect how it is constructed.

Types of Foundations

Houses usually have slab, crawl-space, or basement foundations. Several factors determine which type of foundation is selected for a particular house. Some of these factors relate to site conditions, including the slope of the lot, the location of the water table, or the condition of the soil. In addition, local building codes, the design of the house, buyers' preferences, and even local customs may determine the type of foundation selected. Regardless of which foundation is chosen, builders categorize slab, crawl-space, and basement foundations by the type and amount of space between the floor and underlying soil.

Slab Foundation. For a slab foundation, the underlying soil supports a flat layer of concrete, usually 4 inches thick. Drawing 2.3 shows a slab floor surrounded by a concrete block foundation and a perimeter footing. The foundation wall can also be made of concrete or pressure-treated wood. A monolithic slab, on the other hand, is surrounded by an integral footing poured along with the slab. With this approach, the slab is both the foundation and the first floor of the house. As Chapter 1 notes, a slab home typically requires a nearly flat lot.

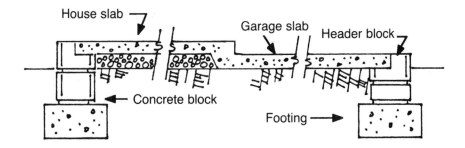

2.3 Slab Foundation

Crawl-Space Foundation. A crawl-space home typically has a perimeter foundation wall of brick, block, stone, poured concrete, or occasionally pressure-treated wood. Interior pillars arranged within the foundation area support the main floor of the house, which is usually made of wood.

The wood floor is located at least 2 feet above the underlying soil, and as Chapter 1 described, the space between the floor and soil is called the crawl space. Builders typically use this crawl space for insulating the floor, running heating and cooling ducts, and installing electrical and plumbing lines. Foundation vents provide air flow to the crawl space and one or more access doors allow for easy maintenance. The builder can build a crawl-space home on a flat lot or one with a moderate slope.

Basement Foundation. A basement foundation combines the best features of slab and crawl-space foundations. With a basement foundation, the bottom floor is normally a slab. Masonry, poured concrete, or pressure-treated wood walls support the main level of the house. These walls rise 8 or more feet high and allow the basement to serve as a functional storage or living area. The underground slab floor is then placed inside the bottom of the walls to keep the walls from moving inward.

The basement walls not only support the house but also need to withstand a significant difference in pressure between the outside soil and the inside basement living area. So, the builder must follow local building codes or work with a designer or engineer to determine what reinforcement may be required for these walls.

The builder next places perforated plastic or clay drain pipes around the base of the foundation near ground level. (See drawing 2.4.) The pipes slope away from the foundation to a storm drain, to the ground surface, or to a sump pump located below the basement floor.

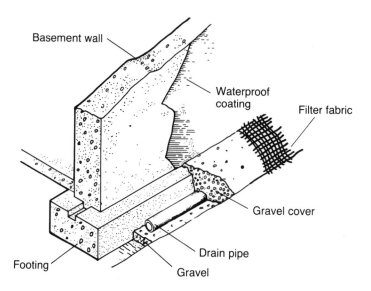

2.4 *Drain Pipe at Base of Outer Foundation Wall*

Since basement foundation walls are built underground but above the water table, the builder must thoroughly waterproof them and, in some climates, insulate them. The builder should coat the walls with a waterproofing mastic or membrane and then place a porous medium, such as gravel, around the wall extending from the ground surface down to the footing drain.

Foundation Materials

Builders construct foundations from three basic types of materials: poured concrete, concrete block, or pressure-treated wood.

Poured Concrete. Poured concrete is the material of choice for slab-on-grade construction. Drawing 2.5 shows a close-up of a monolithic concrete slab. Here the builder has poured the floor slab and footings in one continuous process. This particular slab has a gravel underlayment for extra support, which is optional where the subsoil is sand or gravel and the water table is low.

Before pouring the slab, the builder usually has a licensed pest control company treat the underlying soil for termites. The builder may place some type of vapor barrier—usually a layer of polyethylene plastic—over the soil or gravel. The builder can then add reinforcing rods to strengthen the slab as necessary.

Drawing 2.5 also shows an optional layer of rigid slab insulation that the builder may apply to the exterior of the footing in cold climates. The builder can then erect the main

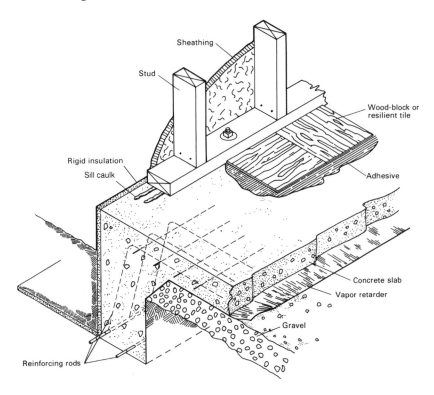

2.5 Monolithic Slab Foundation

walls of the home directly on the slab. In most areas, the bottom plates of the walls, which are the 2x4s that touch the slab, should be made of pressure-treated material. Workers can then apply finish floorings directly to the slab, including carpet, vinyl, or wood.

Poured concrete is also a common foundation material for crawl-space or basement walls. Drawing 2.6 shows a foundation wall being constructed of poured concrete. Once the footing concrete is poured and before it sets, the workers may form a small keyway around the top of the footing so the wall concrete grips the footing.

After the footing sets, the workers use the footing as a base to erect temporary forms to hold the wet concrete. The sketch shows timber and plywood forms, which builders sometimes use. However, modern metal forms are rapidly replacing the traditional wood forms. After all of the forms are in place,

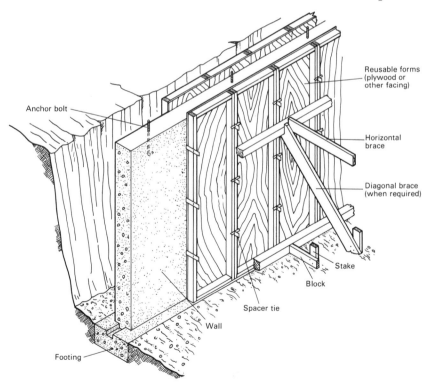

Anchor bolt

Reusable forms
(plywood or
other facing)

Horizontal
brace

Diagonal brace
(when required)

Stake

Block

Spacer tie

Wall

Footing

2.6 Poured-Concrete Foundation Wall

workers install the reinforcing and pour the concrete from the top between the two sides of the forms. Before this concrete sets, the builder usually inserts anchor bolts into the walls.

Concrete Block. Builders also use concrete block or sometimes brick for basement or crawl-space foundation walls. Drawing 2.7 shows a section of a concrete-block basement wall. Notice that the builder has inserted a window for light and ventilation in the basement. Below the grade, workers have applied a coating of cement parging, which is mortar cement used for waterproofing, and other waterproofing material.

In those parts of the country with colder climates, the builder may apply both waterproofing and insulating materials to the basement walls. The sketch shows that the coating is heavier near the bottom and slanted outward to help move

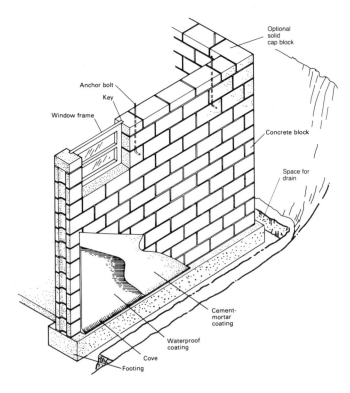

2.7 *Concrete-Block Foundation Wall*

moisture away from the base of the foundation after the soil is backfilled against the wall. The top of the wall shows anchor bolts and optional solid cap blocks, which building codes may require in some areas.

Pressure-Treated Wood. Builders sometimes use pressure-treated lumber for basement and crawl-space foundations because wood walls are easier to insulate, wire, and finish than block or concrete walls. For this type of foundation, the wood is treated with a preservative to resist insects and rot.

Drawing 2.8 shows the installation of a typical pressure-treated wood foundation wall. In this situation, workers construct the foundation on a layer of gravel placed below the frost line. This layer of gravel serves as both a footing and a level bed below the basement floor slab. The gravel also assists in carrying moisture away from the floor to any drainage system outside the house.

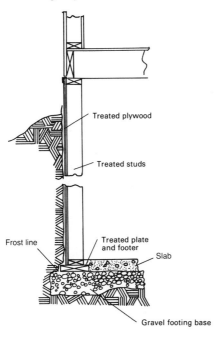

Treated plywood

Treated studs

Frost line

Treated plate and footer

Slab

Gravel footing base

2.8 Pressure-Treated Wood Foundation Wall

Treated wood foundation walls are often preconstructed as panels at a factory and shipped to the site for assembly. Generally the panels consist of a frame of treated 2x4s, 2x6s, 2x8s, or 2x10s spaced according to the load they will support. Workers nail a treated plywood facing to the frame, caulk the joints, and wrap the entire exterior with polyethylene sheeting moisture retarder. Builders have successfully used pressure-treated wood foundation walls for basements in many parts of the country.

Foundation Design

Why would a builder recommend one foundation design or material over another? First, local soil and terrain conditions may require the use of certain foundation designs or materials. As already mentioned, in certain parts of the country, expansive soils make the construction of crawl-space or basement foundations difficult. In such situations, a specially designed basement or slab foundation may be needed. As another example, a basement may be a less expensive approach for a sloped site. On the other hand, construction of a basement on a site with underlying rocks may be more expensive because of the cost of blasting and removing the rock.

Regional geological conditions and climate may also influence the design of the foundation. For example, homes built in areas that are subject to earthquakes may require heavily reinforced foundations. As already mentioned, cold weather may require special techniques for pouring and curing concrete. In addition, because workers cannot mix and apply mortar to blocks and bricks on extremely cold days, the builder may consider using a wood foundation in colder climates.

The cost and availability of material may also influence the design of the foundation. As one example, clay bricks are only made in certain parts of the country. Shipping them from other areas can be expensive. If a material or method is uncommon to a particular area, the builder may also have difficulty finding a tradesperson who knows how to work with that material, which can again increase construction costs.

The overall design of a home may also dictate the choice of foundation material. Whether using stock plans or plans designed by an architect or designer, the builder must pay careful attention to the details of those plans. If the house plans call for a basement, then building a crawl-space or slab-on-grade foundation can result in a very different style of house. For example, if a builder adds a basement, the plans will have to be modified to accommodate the basement stairs and other changes.

In selecting materials and a design for the foundation, the builder also needs to pay close attention to local customs. If a neighborhood is filled with crawl-space homes, then a builder may prefer to build a home with a crawl space in that area. If no one else in the area has a basement, then there is probably a good reason for that custom. Besides any practical reasons for the construction of particular types of foundations, radical departures from local customs can sometimes adversely affect the resale value of the home.

Constructing the Rest of the Foundation

Even after extensive planning, constructing the rest of the foundation after the footings entails more than just stacking up blocks and mortar. The builder must make sure the structure's foundation is perfectly level. Since the foundation is built on top of the footings, the masons or concrete formers must correct any inconsistencies in the footings as they construct the rest of the foundation. The masons have to follow the batterboard lines carefully, and a good chief mason will recheck the lines repeatedly to ensure that the foundation is level and square.

At this time, the masons, concrete formers, or other tradespeople working on the home may also install supporting material and other nonmasonry items in the foundation. These items may include reinforcing, as well as—

- crawl space vents and basement windows
- posts, beams, and girders for supporting the main structure
- lintels, which are lengths of angled steel that support any masonry over the top of a window, door, or other opening
- foundation drains and a sump pump if the foundation drains do not drain to the top of the ground
- foundation insulation
- cement parging on concrete block

- dampproofing with an asphalt coating or polyethylene film
- stucco or other decorative coating above grade

Before a slab or basement floor is poured, a plumber installs any needed water and sewer lines. In some areas, the builder may also need to install a pipe through the slab into the gravel under the slab for venting radon soil gas into the outside air.

As the next step in constructing the foundation, the workers pour the concrete floor slab. They then level it to a rough finish and, when it begins to set, they trowel the concrete into a smooth, hard surface. These two steps are necessary for a strong, lasting concrete slab. However, using too much water can cause the aggregate—the stones or rocks contained in the concrete—to fall to the bottom. This weakens the slab and softens the wearing surface. Here again, the builder can ensure that a trained professional has the experience to deliver a well-built concrete structure.

Building codes require large bolts in the concrete foundation or slab, which extend up so the walls or floors can be bolted down. The builder can install these bolts as the concrete is poured or later by drilling holes at the appropriate places and securing the bolts with anchoring devices. Again, local codes, customs, and traditions will dictate how builders install anchor bolts in the foundation or slab.

In addition to wall thicknesses and construction methods, local building codes dictate the size and number of basement windows or crawl-space vents. The builder will know how to determine the proper number of windows and vents for each home in a particular area based on local codes.

In some parts of the country, basement and foundation insulation is a critical energy-saving feature. The most common method is to install fire-resistant fiberglass insulation on the inside of the basement wall. Another method is to install a rigid insulating board on the outside of the foundation, along with the waterproofing, before the backfill is replaced in the trench around the foundation.

Crawl-space foundation walls are usually not insulated, since in most parts of the country the builder installs insulating batts between the wood floor joists for this type of house.

Also, with a crawl-space foundation, a builder usually applies a layer of polyethylene directly on the soil within the crawl space area to reduce moisture and the possibility of wood damage from fungus or rot.

The type, material, and design of a foundation depend on many factors, from building codes and site conditions to buyers' preferences and local customs. Builders generally construct foundations from either poured concrete or concrete blocks. However, builders also sometimes use pressure-treated wood foundations because they are easier to insulate and finish on the interior.

As the builder places the slab, crawl-space, or basement foundation on the footings, it must be level and square. Once the concrete has properly cured and all foundation supporting materials are in place, the builder can then proceed to the next stages of house construction—the framing, roofing, and siding.

3

Framing, Roofing, and Siding

The rough carpentry phase of house construction may look simple enough. After all, you may have built a dog house or a bird feeder when you were young. In reality, however, framing, roofing, and siding are highly technical fields with skills, terms, and procedures that must be mastered to build a sound home.

Framing

A builder may select a particular framing approach from a variety of effective techniques. However, this chapter will explain the rough carpentry phases of constructing a house in general terms without emphasizing any particular technique.

Planning Materials and Techniques

Long before construction starts, the builder plans the framing of the house. Since the framing is anchored to the foundation, the builder must plan these two parts of the house together. Planning the details of framing is also part of the design process. For example, such things as the planned roof line and the location of the interior walls and stairwells will affect the plans for framing.

Notice the floor framing in drawing 3.1, especially the floor joists. A joist is a beam constructed of dimension lumber—often 2x8 or 2x10—that provides direct support to either a floor or ceiling. Underneath these floor joists is a support beam held up by piers.

The builder and designer determine the size of the joists and support beam and the size and spacing of the piers according to the amount of weight that the homeowner will place on the floor. Living spaces occupied by furniture and people require a structure of moderate strength. On the other hand, bathrooms with heavy bathtubs may require a stronger structure.

The wall framing must take into account the size and placement of windows and doors, roof loads, and support beams. The center beam in drawing 3.1 supports a center wall that in turn braces the roof. If the builder uses a trussed roof instead, this center wall might be unnecessary. As discussed later in the chapter, a trussed roof has preassembled components that provide support for the roof.

Once the floor and wall framing is planned, the builder plans the ceiling and roof framing together. This is because the roof rafters, or wood beams that support the roof, and the ceiling joists form an integrated roof system.

Notice the wall studs in drawing 3.1. They are made of slender wood or metal columns, usually 2x4s, placed vertically in the wall. Horizontal plates of similar material connect the studs at the top and bottom. Diagonal metal or wood bracings are often used to brace the wall assembly. The builder places headers—beams made of doubled pieces of dimension lumber—in load-bearing walls above the doors and windows. (See drawing 3.1.) The header carries the load of the roof or upstairs floor above the window opening to the studs on either side of the opening.

Floor Framing

Once the builder has finished planning the materials and techniques for framing, the next step is to begin construction

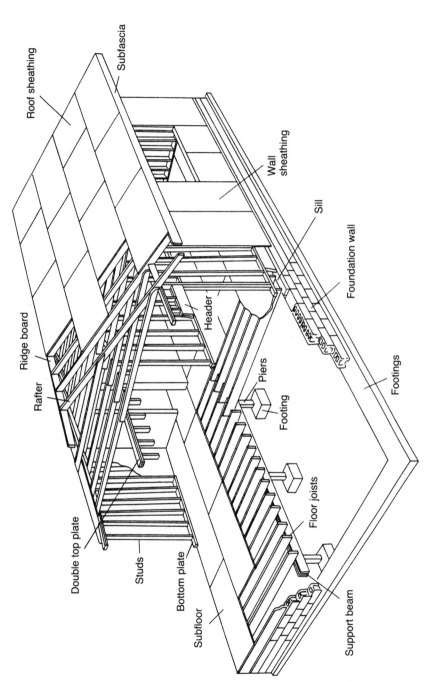

Roof sheathing

Subfascia

Wall sheathing

Sill

Foundation wall

Footings

Footing

Piers

Header

Ridge board

Rafter

Floor joists

Support beam

Double top plate

Studs

Bottom plate

Subfloor

3.1 Cross Section of House Framing

of the floor. Drawing 3.2 shows the bare essentials of floor framing. When framing the floor in most climates, the builder simply nails the joists to a sill, which is bolted to the foundation. In high-wind or coastal areas subject to strong storms, however, local codes may require builders to fasten the floor and the wall above directly to the foundation with some anchoring system.

Typically, the builder places the joists on the sill plate of the foundation and on a beam or wall under the center of the structure. The lumber industry provides builders with standards for the strength of a given piece of dimension lumber. From these standards, the builder knows how many feet a floor can span from one support to another. The builder can also determine the spacing of piers needed to support the beam that in turn supports the joists.

The builder usually laps and nails the joists together in the center. (See drawing 3.2.) If the builder plans to place a load-bearing wall on this floor running parallel to and between the joists, then the builder may use blocking to carry the load to the adjacent joists on either side of the wall. If a load-bearing wall rests directly on a joist, then the builder may double or triple the joist, depending on the load.

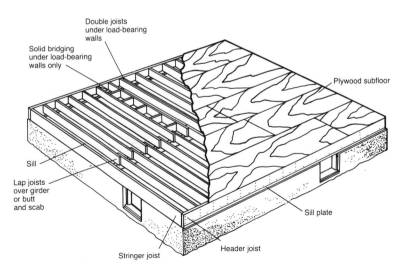

3.2 Floor Framing

With proper planning at the design stage, the builder can avoid beams or columns in large living areas. For example, the builder can use a steel or wood-laminated beam to extend the area beneath a floor or roof. In other words, the stronger the beam, the larger is the open area that can be created without supporting columns. In addition, to avoid having beams project below the ceiling level, the builder can attach the joists to the side of the beam to make the ceiling flush. (See drawing 3.3.)

The builder then attaches a subfloor to the floor joists. (See drawing 3.2.) This may be a single layer of material or two thinner layers. For subflooring in today's homes, builders use plywood, waferboard, or oriented strand board and nail

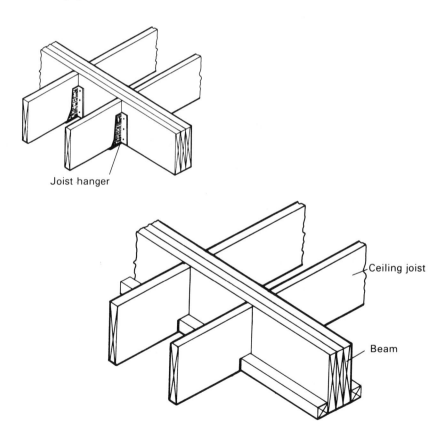

Joist hanger

Ceiling joist

Beam

3.3 Framing of Flush Ceiling Beam

or glue this material to the joists. Once the subfloor is in place, the builder can proceed with the wall framing.

Wall Framing

Several carpenters usually work together to construct the walls of a house, which are typically framed with 2x4 lumber. As the first step, the carpenters use a chalk line to mark the location of all walls on the subfloor. They next lay the top and bottom plates on the subfloor and mark where the studs go on the plates. (See drawing 3.4.)

The carpenters then arrange the studs between the plates. At this point, they also make openings for windows and doors by placing headers above the openings and placing extra studs on either side to carry the load over the openings. Once the carpenters have the bottom plate, studs, and top

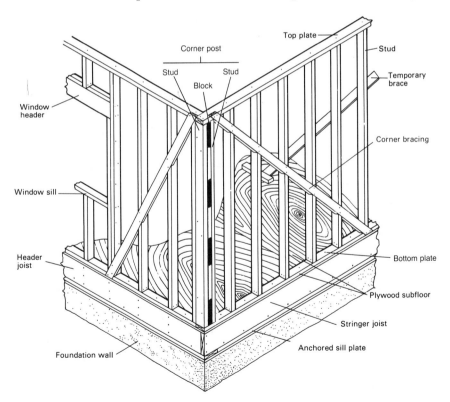

3.4 Wall Framing with Platform Construction

plate in place, they nail this system together as it lies on the floor. (See drawing 3.4.)

Typically, the carpenters nail special corner posts in place where walls form corners. With the traditional corner post, the carpenters nail two studs together with blocks in between. (See drawing 3.4.) The carpenters then place these two studs at the end of one wall and a single stud at the end of the other wall. That way, they have sufficient wood to nail the walls together and adequate surface for fastening the drywall to the inside of the walls. More recently, many builders have begun using metal clips at corners for drywall support, thereby eliminating the need for one of the studs and the blocks. Either method is acceptable.

Before the wall is lifted into place, a carpenter attaches either wood or metal diagonal bracing near the corners of the structure. Figure 3.4 shows a 1x4 wood brace that a carpenter has attached to the frame by cutting diagonal notches in the studs. However, many builders prefer to use metal braces, which do not require notching.

The carpenters must carefully measure and attach the braces to ensure that the wall frame is perfectly square—or forming a perfect rectangle—and plumb or vertical when it is lifted into place. The lead carpenter also checks to make sure the window, door, and other openings are the correct size and in the location called for in the house plans. (See photo 3.5.)

After the wall is nailed together, the carpenters lift it into place and support it with temporary bracing. The lead carpenter will use a level to check that the wall is perfectly vertical. The crew next straightens the wall by placing a tight string along the top and moving the wall until the top is in line with the string at all points. The crew then firmly nails the braces down to the floor to hold the wall in place.

The carpenters next nail the exterior wall sections together and usually double the top plates for reinforcement. If the roof framing rests directly over the studs within the exterior walls, the double top plate may not be needed. Once the carpenters erect the walls, they then anchor them down by

3.5 Framed Wall and Window Opening

nailing through the bottom plate and the subfloor to the underlying joists.

To provide room between the studs for additional insulation in colder climates, builders increasingly use 2x6 studs and plates for exterior walls. Also 2x6s are frequently used for walls that will contain bathroom or kitchen plumbing lines. In addition, when placing the wall, a carpenter may apply caulk under or around the bottom interior edge of the exterior walls where they join the subfloor. This step can reduce air leakage. In high wind areas, the builder may nail hurricane straps to the exterior of the wall and firmly attach the studs to the underlying structure.

The builder may also apply sheets of sheathing to the exterior walls. (See drawing 3.1.) This sheathing both reinforces the wall and provides a base for the external finish material. For this sheathing, the builder may use a rigid foam insulation, often backed with aluminum foil. Over the top of this, the builder may then apply an air infiltration barrier to reduce air leakage.

Constructing the walls is labor intensive and subject to the tyrannies of weather. For this reason, many builders construct wall panels in a workshop, where tools can be conveniently put away each evening, construction can go on in bad weather, and workers can even finish tasks at night during rush times. With this approach, builders can also avoid delays by keeping a ready supply of raw materials on hand in the workshop. They can even use scraps of material left over from one job for other jobs, so no materials are wasted.

If panels are constructed in a workshop, the builder loads them onto a flatbed truck, not unlike the truck that would have hauled the raw materials out to the job site. These wall panels are essentially identical to the walls built on site. However, such innovations can save on construction costs and thus contribute to a more affordable home.

Ceiling and Roof Framing

Traditionally, the builder frames the ceiling and roof on site with dimension lumber like the floor system. As noted in the Introduction, this process is sometimes called stick-built construction.

Stick-Built Roof Framing. After the walls are up, the carpenters cut and lift the ceiling joists into place. These are typically made of 2x6s or 2x8s, usually spaced 16 inches apart. The ceiling joists, which are usually planned to run parallel to the roof rafters, hold the walls together and support the ceiling.

The roof framing also includes wood rafters. (See drawing 3.1.) In a typical roof system, one end of each rafter rests on an exterior wall and a portion of the rafter may hang over the edge of the wall to provide an overhang at the eaves. The other end of the rafter slopes up to meet another rafter coming from the opposite side of the house. The rafters meet in a ridge and the carpenters then nail the rafters to a ridge rafter.

About a third of the way down from the ridge to the ceiling joists, the carpenters nail collar beam braces horizontally between opposing rafters. They usually install one collar beam for every three or four rafters as required by the local building code. In a plain gable roof, the rafters and ceiling joists form an open triangular space that becomes the attic of the house. A 2x4 gable end wall encloses the ends of the attic. (For a picture of different types of pitched roofs, see drawing 3.6.)

Most new homes call for attic ventilation. Some home designs use a vent installed within the gable end wall, which may be both decorative and functional. Drawing 3.7 shows four different designs for vents (A, C, D, and E) and a cross section of a vent (B), which shows how the vent is constructed. Other alternatives include soffit, roof, or ridge vents (F).

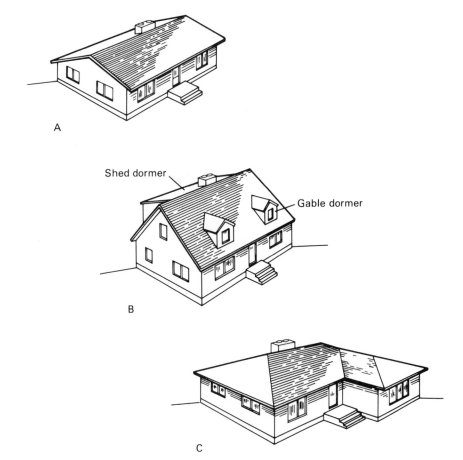

3.6 Pitched Roof Types: (A) Gable, (B) Gable with Dormers, and (C) Hip

After carpenters have completed the gable end walls and put all permanent bracing in place, they apply the roof sheathing, which is usually plywood, waferboard, or oriented strand board. (See drawing 3.1.) They install the long dimension of the sheathing perpendicular to the rafters and

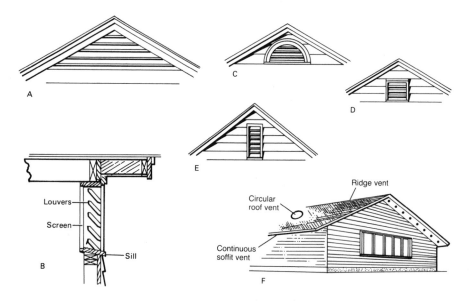

3.7 Vents: (A) Triangular Gable Vent; (B) Gable Vent Cross Section; (C) Half-Circle Gable Vent; (D) Square Gable Vent; (E) Vertical Gable Vent; and (F) Soffit, Circular Roof, and Ridge Vents

usually stagger the sheathing to provide extra bracing. The carpenters may extend the sheathing over the gable ends to form a gable end overhang.

Wood-Truss Roof Framing. Many builders construct roofs with preassembled components called trusses, in which rafters, ceiling joists, and intermediate braces are fastened together with steel plates. In fact, roof trusses have become so common that many cities have specialized roof-truss manufacturers.

Roof-truss manufacturers can produce a custom roof-framing system for a given set of plans. The roof-truss system replaces a roof constructed on site with ceiling joists, rafters, collar beams, ridge boards, and gable-end framing. Since a structural engineer custom designs the truss system, usually with the aid of specialized computer programs, the builder can use smaller dimensioned lumber to achieve significant savings for the buyer.

Trussed roofs are engineered to span the distance between the front and back walls of the house. Since the trusses only bear on exterior walls, the builder can move interior walls around with great flexibility. This allows for the construction of large rooms that can run the entire width or depth of the house.

Wood trusses are also normally spaced farther apart than stick-built rafters, which again saves on lumber costs. However, trusses generally only gain such savings with simple roof lines. When a roof is too complex, the truss design can become so complicated that the savings may be lost.

In addition, fully vaulted ceilings are impossible to achieve with trusses. However, builders do use scissors trusses with sloping bottom members to support partially vaulted ceilings. Overall, the combination of the cost savings and the added design flexibility makes roof-truss framing an attractive way to frame a roof.

Drawing 3.8 shows a simple sketch of trusses being erected. In the first step, the builder lifts the gable truss into place, often with a crane. Care is taken to make sure that this truss is perfectly positioned and exactly vertical. As indicated in the sketch, the builder firmly braces the gable truss. The builder then lifts the other roof trusses in place and temporarily braces them to this gable truss. A trussed roof does not need a ridge board. Instead the builder installs 2x4 blocks between the trusses or ceiling joists and later nails the tops of parallel walls to these blocks.

Roofing

After the roof is sheathed, the carpenters usually tack in place a layer of asphalt-treated felt paper. (See photo 3.9.) As part of the roofing material, this paper protects the roof sheathing from rain and sun until the roofers can apply the shingles.

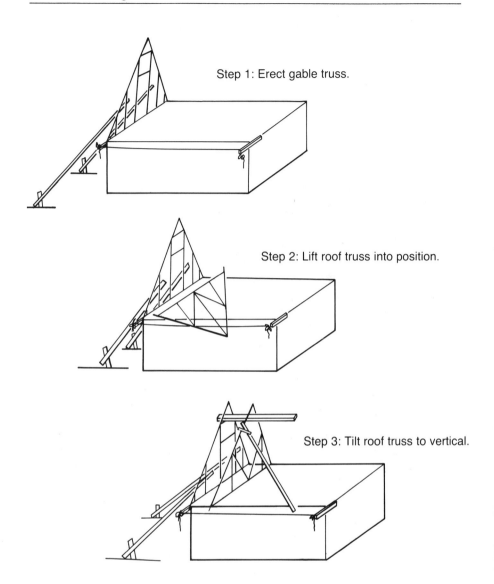

Step 1: Erect gable truss.

Step 2: Lift roof truss into position.

Step 3: Tilt roof truss to vertical.

3.8 Erection of Roof Trusses

Roofing Material

Roof coverings come in many different materials, textures, sizes, shapes, colors, and application styles. The choice of roof covering for a home depends on such factors as local traditions, climate, construction budget, availability of

3.9 Newly Sheathed Roof Covered with Felt Paper

skilled roofers in the area, and the tastes of the homebuyers and designers.

For example, terra-cotta tile roofing may be popular in California, Florida, or Arizona, where the Spanish tradition and local climate make this an appropriate material to use. However, finding a skilled tile roofing crew in Virginia or Pennsylvania might be difficult, where asphalt shingles, wood shakes, or wood shingles might fit better with local tastes and customs. (See drawing 3.10.)

Another roofing material that builders use is slate, but this is usually limited to expensive custom homes. Tar and gravel or other forms of membrane roofing are used for flat or built-up roofs. Sheet metal roofs made of aluminum or galvanized steel are also found in some rural areas.

Builders still use copper for some roof parts, particularly for flashing and roofing at bay or dormer windows. In addition, sheet metal, such as aluminum or galvanized steel, is often used for edging and valley or other types of flashing. (See drawing 3.11.) Builders place flashing where the roof meets the walls or chimney to keep water from entering the house.

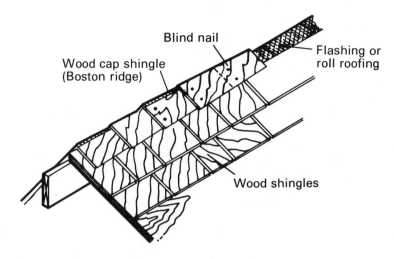

Blind nail

Wood cap shingle
(Boston ridge)

Flashing or
roll roofing

Wood shingles

3.10 Wood Shingles

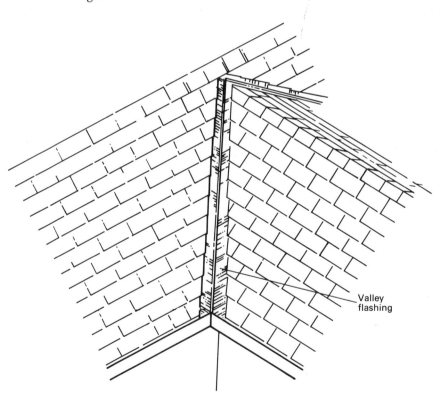

Valley
flashing

3.11 Valley Flashing for a Roof

The most common roofing material is a 3-foot-wide, asphalt-fiberglass shingle. These shingles are sold in bundles. Three bundles cover 100 square feet of roof and are called a square. A typical square of fiberglass roof shingles weighs over 200 pounds. Since the roof of a typical three-bedroom ranch house can use 20 to 30 squares of shingles, it is easy to see why the bracing and engineering of the roof are so critical. Since a typical bundle of shingles weighs 80 pounds and must often be carried by hand up a ladder to the roof, it is also easy to see why roofing requires skill, experience, and strength!

Roofing Process

Once the felt paper is in place, the roofers can begin their job. To ensure proper installation of the shingles, the roofers place marks at the gable edges of the roof for the correct spacing of the shingles and then snap chalk lines between them to mark where they will nail the shingles.

If called for by the house plans, the roofers may nail a metal drip edge to the bottom edge of the roof sheathing. As an alternative, the roofers may simply let the shingles hang over the eaves a bit to provide this protection. On steep or high roofs, special safety precautions are required during the roofing process.

The roofers nail a starter strip of shingles on the bottom edge of the roof. Subsequent rows of shingles overlap each other to form a double or triple layer everywhere. Beginning from the lowest edge of the roof and keeping the tabs in line vertically and horizontally, the roofers then nail shingles in staggered rows up the roof. They continue this procedure up both sides of the roof until the shingles meet at the ridge.

The roofers may then install a ridge vent. Alternatively, if gable vents are used, the roofers cut cap shingles that they then nail over the ridge and extend down either side of the roof. If wood shingles or shakes are used, the roofers nail a special roll roofing or flashing over the upper rows of shingles or shakes to prevent water leakage. (See drawing 3.10.) They then cover the flashing with ridge shingles or shakes.

As the roofers install the shingles, they seal any vents that extend through the roof with a special collar flashing, which is premanufactured to the specific size needed. The collar flashing is fashioned with a flat base that lies on the roof and a collar that fits tightly around the vent pipe at the top. The upper side of the base laps under the shingles above the vent, and the lower side of the base laps over the shingles below the vent. The roofers then seal the collar with durable caulk or black plastic cement.

As shown in drawing 3.11, when two roof lines meet and form a valley, the roofers must take special care to make sure the valley does not leak. One approach is to cut the shingles away and form an exposed valley that is protected under the shingles with flashing made of copper, aluminum, or galvanized steel.

A more recent method is to install roll roofing in the valley and then to weave shingles from adjoining roof lines into each other to form a continuous covering of overlapping shingles over the valley. Water is then carried away over the shingles with little risk of leakage. However, this approach requires the skills of an experienced roofer. Generally, local customs dictate the method of flashing that a builder may use.

Roofers also use flashing where a roof abuts a wall. (See drawing and photo 3.12.) If the wall is covered with wood, vinyl, aluminum, or some other type of siding, then the roofers nail short pieces of step flashing on top of each course of shingles and bent up under the siding, with most of the flashing covered by the succeeding row of shingles.

If the wall is brick or where the roof meets a brick chimney, then the roofers install additional pieces of flashing in the mortar joints, which fold down to cover the top edges of the step flashing. They then caulk the joints between the flashing and brick wall.

On the upper side of chimneys where the roof abuts the chimney from above, the roofers construct a saddle or cricket, which is a tent-shaped projection that directs water around the chimney. A large saddle or cricket may have a

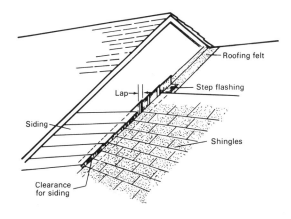

3.12 Flashing at Roof-Wall Intersection

ridge board, short rafters, and plywood sheathing to support the flashing.

Exterior Trim

Once the roofing is completed, the builder can then turn to work on the exterior trim. Most homes have some sort of eave overhang or soffit, which is a portion of the roof extending out over the side of the house. While a soffit appears decorative on most styles of homes, it serves the important function of carrying water away from the walls and foundation. In older homes in the South, a broad overhang on the south side of the house also adds shade to the windows and doors and helps to cool the home in the summer.

Drawing 3.13 shows a portion of an eave or soffit that runs parallel to the ground at an exterior wall. Portions of soffit that run at an angle to the ground at a gable end wall are called rake overhangs. In drawing 3.13 a portion of the roof truss forms the underlying structure of the eave overhang.

At the ends of the rafters or trusses, the workers nail a fascia board—often a 1x6—to cover the gap between the sheathing and the soffit. The fascia board may have a groove cut into it lengthwise to receive the soffit. For easier maintenance, the fascia may also be made of prefinished aluminum or vinyl. The builder usually installs the fascia before the soffit because especially with aluminum or vinyl soffit the fascia supports the soffit.

The bottom of the truss structure, which drawing 3.13 shows as a short 2x4 that runs parallel to the ground, is called a return. Literally, this 2x4 returns from the end of the truss to the side of the house. In fact, it should extend to the sheathing that the carpenters apply when they frame the walls. To the bottom of this return workers nail plywood, aluminum, vinyl, or other material to enclose the soffit. The builder may then install soffit vents to openings in the plywood or else install perforated sections of aluminum or vinyl for ventilation.

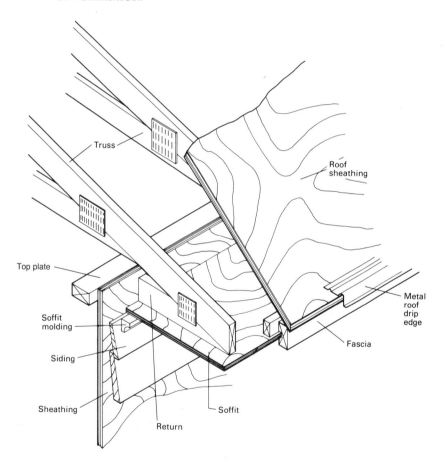

3.13 *Eave for a Truss Roof*

After the workers apply the siding, they then nail a continuous strip of molding to the wall directly under the soffit to cover the gap between the top of the siding and the soffit and to provide additional support for the soffit. This molding is called the soffit molding. Aluminum or vinyl channels are generally used to support soffits made of these materials.

If the builder applies brick on the walls, a frieze board (not shown in drawing 3.13) is often used at the top of the wall under the soffit. A frieze board is typically a 1x4 or 1x6 board and may have a decorative molding where it meets the soffit.

Exterior Windows and Doors

After builders have framed and sheathed the walls and completed the roofing but before they have installed the exterior siding, they can install the doors and windows.

Windows

Windows do three things: they admit light, provide a view, and give ventilation. Most windows installed today have double- or even triple-insulated panes. (See photo 3.14.) A vacuum or special gas in the gap between the panes reduces heat flow through the sealed unit. Low-E coatings on the glass can also increase energy efficiency at little added cost.

Unlike small, inexpensive panes used in older, uninsulated windows, insulated panes are expensive and usually made to the full size of the window. Muntins, the wood trim pieces that divide traditional windows into small sections, actually provide support in older, uninsulated windows. However, muntins are just decorative pop-in pieces in newer, insulated models.

Window Styles. Many home designers make good use of the latest varieties and innovations in window styles to enhance both the appearance and functional value of the home. The most common type of window in traditional homes is the double-hung window. (See drawing 3.15.) The most common frame material for windows is still wood, although vinyl, metal, or combinations of these three materials are also common. In the more exotic models, the sashes

3.14 Window Installation

pop out or tilt in for easy cleaning. Also, window screens are often built into the window unit.

Another common type of window, particularly in more contemporary homes, is the casement window, which usually swings out by using a hand-cranked gear or lever. An awning window is essentially a casement window built to open horizontally. (See drawing 3.15.) Electric openers and remote controls are available to open high windows. In fact, coupled with a remote-controlled electric opener, an awning window can become a handy ventilation device in the peak of a vaulted ceiling.

Stationary or fixed windows are available in rectangular picture windows (see drawing 3.15), as well as a variety of decorative shapes. Triangular or trapezoidal windows can provide light from triangular gable walls in contemporary homes with vaulted ceilings. In addition, many traditional

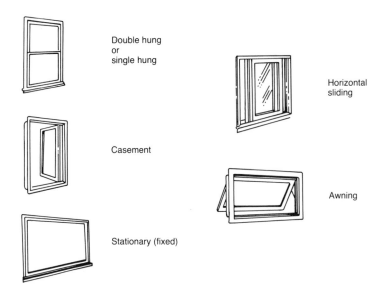

Double hung
or
single hung

Horizontal
sliding

Casement

Awning

Stationary (fixed)

3.15 Common Window Styles

styles of homes have featured round or octagonal windows for centuries.

Finally, some builders use horizontal sliding windows, sometimes combined with a fixed panel. The most common version of this type of window is the sliding glass door, which provides a broad expanse of light to open up an otherwise small room and offers access to and from the house.

Window Selection. The choice of window style affects the appeal, value, and livability of a new home. When selecting the appropriate style, material, and type of window for a home, the builder, designer, and even a window specialist may meet to discuss the full range of options. As part of that process, they may consider the following factors:

- **style of the home**—The choice of window style should fit into the design of the home. For example, builders typically use double-hung windows with muntins in traditional styles of homes, while casement, awning, or horizontal sliding windows appear more commonly in contemporary homes.

- **budget**—Most suppliers have several styles of windows available at affordable prices, with more expensive windows offering many additional features.
- **materials**—Wood, aluminum, vinyl, vinyl-covered wood, or fiberglass-reinforced plastic windows go with either contemporary or more traditional homes.
- **maintenance**—The builder should consider maintenance when selecting windows. For example, where the budget allows, the builder might use one of the tilt-in models that aid in cleaning the exterior surface of the window.
- **manufacturer**—In selecting windows, the builder also considers the reputation of the manufacturer to make sure that the windows come from a reputable source.
- **insulation, light, and air**—Finally, in selecting windows, the builder will also consider their energy efficiency. In northern regions or where air conditioning is used, finding windows that prevent air leakage and provide superior insulation may top the builder's agenda. In milder climates, lighting and ventilation may become more important.

Doors

Once the builder has chosen the windows, the next step is to select the doors. Like windows, doors come in many types, styles, and materials. In fact, the same manufacturer that provides the windows may also provide the doors and related trim materials. (See photo 3.16.) Again, the house plans largely dictate the style of doors and trim that the builder uses. For example, a builder would avoid colonial trim for the door frames of a contemporary home. Simi-

3.16 Premanufactured Door and Exterior Trim

larly, high-tech door styles would look out of place for the entrance of a colonial home.

Some manufacturers offer fiberglass or steel doors in a wide variety of styles, including traditional and contemporary. These doors are usually filled with foam to conserve energy and include other energy conservation features, such as insulated glass and magnetic weatherstripping. Some fiberglass doors are even stainable.

Once the exterior doors and windows are selected and the windows or at least the window frames are installed, the builder can next install the siding.

Siding Material

Buyers' preferences, climate, local customs, availability of material, construction budgets, and the design of the house determine a builder's choice of siding material. In many areas, home builders have traditionally used a wide array of wood material for siding, which they can finish naturally, paint, or stain. Some types of wood-based products, such as hardboard siding, are also used for siding material. In addition, brick, stone, or stucco siding are common in many areas of the country, while vinyl, steel, and aluminum siding are also popular because of the lower maintenance and cost.

However, this list is in no way complete. Builders may use wood shingles, adobe, decorative concrete block, textured poured concrete, fiberglass, metal, and even more exotic materials for siding. Some new homeowners have even chosen to build their homes into the sides of cliffs where they use packed dirt mixed with Portland cement as siding material. However, brick and stone, stucco, and wood are still some of the most common siding materials used on new homes today.

Brick and Stone

Brick and stone are common siding materials in the East, Midwest, and South. They provide the ultimate in durability and low maintenance but generally are more costly to install. Additionally, the type of stone or specific color of brick the builder can use depends on the location and availability of the material.

Your author wanted to duplicate a special brick color from bricks made in the 1940s for a remodeling project. Unfortunately, while the manufacturer was still in business and the style of brick was still available, the special color came from clay found only in one quarry, which is now covered by a lake.

Planning for a brick exterior, called brick veneer when installed over a frame wall, must begin early. With this type of exterior, the builder constructs the foundation with a ledge at the finish grade to accommodate the brick exterior surface. In addition, brick and stone are porous materials. If these materials are used for siding on a home, the builder should place asphalt-treated felt over the sheathing before laying the brick and should install weep holes near the bottom of the wall.

Stucco

Popular for many years in many parts of the country, traditional stucco is a cement-based product installed in several coats over a metal lath that is nailed to the framed wall. (See photo 3.17.) The final coat sometimes has a pigment that gives the wall its color in place of an additional coat of paint. Recent advances have allowed builders to use synthetic materials to give homes that distinctive stucco-like finish.

Wood

Builders prefer to use wood or wood-based materials that are easily painted, easy to work with, and relatively free from warpage and splits. Common woods that have these qualities include cedar, several varieties of pine, and cypress. Builders also use western hemlock, Douglas fir, spruce, and poplar. (For an example of wood siding, see photo 3.18.)

If the wood is to be painted, the builder selects wood as free from knots as possible. Wood tends to shrink and expand with the seasons and natural weathering of the siding is inevitable, so paint will not last forever. However, an experienced builder can determine what type of wood

3.17 Stucco Application

best suits the climate of a particular area and what type of paint and how many coats to apply. Many wood siding products come with one rough and one smooth side. If the builder plans to stain the wood, then the rough side may be exposed to the exterior. Otherwise, the smooth side is preferable for painting.

One common way to apply wood siding is as horizontal lapped siding. With this type of siding, the builder begins at the lowest point on the wall and nails long strips of wood horizontally upward until reaching the underside of the eave, with each subsequent piece lapping the piece below it.

Builders can also install some types of wood siding in vertical or even diagonal patterns. One method that is popular for some contemporary home styles is to use rough-sawn boards and battens nailed vertically. Battens are narrow strips of wood that cover the vertical joints between the boards. Builders can arrange these boards in different patterns to suit the particular design of the house and tastes of the homebuyers. They can also vertically install and stain some plywood and other types of panel siding material to give a similar effect. In this case, the manufacturer cuts

3.18 Wood Siding

grooves in the panels at the factory to simulate the appearance of boards and battens much less expensively.

Other types of siding include wood fiber products like hardboard or aluminum, steel, or vinyl siding. Builders typically install all of these types of siding horizontally. (For an example of vinyl siding installation, see photo 3.19.) Furthermore, these products usually come preprimed or prefinished to reduce the painter's chores.

Painting and Staining. Once the wood or wood-product siding is installed, the builder often has the siding painted or stained. The builder will also usually have the exposed wood surfaces in brick- or vinyl-sided homes painted. Exterior painting has three basic steps: priming, caulking, and finish painting.

The prime coat of paint protects the surface of the wood from moisture. Particularly in humid climates, the painter primes the wood siding and trim as soon as possible after it is installed.

After priming, the painter caulks all cracks, joints, and seams thoroughly with a caulking material designed to work well with the selected paint. In areas where the

3.19 Vinyl Siding Installation

humidity shifts throughout the year, wood joints expand and contract, and the caulk must be able to withstand these changes.

As the final step, the painter applies the finish coat of paint. For some situations, the painter may recommend two coats of exterior paint. While a bit more expensive, this approach is more durable. Stains may be less durable than a good paint job, but they are easy to reapply and are often preferred for a more natural appearance.

Ventilation

The builder determines the type of attic ventilation early in the home design phase. (See drawing 3.20.) In southern climates, adequate attic ventilation can lower the average attic temperature by as much as 50 degrees during the hot summer months. This reduction in attic temperature can reduce cooling costs in hotter weather. Besides controlling the temperature in the summer, attic ventilation is needed to prevent moisture from accumulating in the winter.

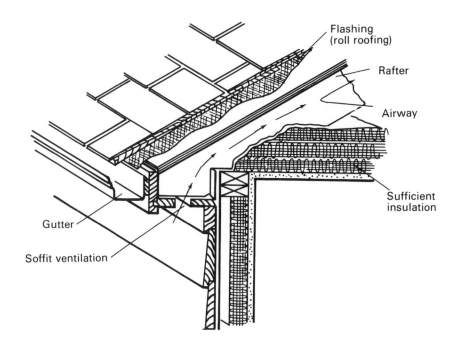

Flashing
(roll roofing)

Rafter

Airway

Sufficient
insulation

Gutter

Soffit ventilation

3.20 Inlet for Attic Ventilation

Builders typically use gable vents (louvers) or a ridge vent and soffit vents for attic ventilation. (See drawing 3.7.) Hot air in the upper most reaches of the attic is carried out of the gable vents or a ridge vent by convection. As the hot air leaves the peak of the attic, cooler air is pulled in through the soffit vents to replace it. Building codes specify the minimum area of ventilation that the builder should provide. If the builder does not install soffit vents, the other vents must be substantially increased.

Homebuyers in the hottest climates may want some type of power ventilation. The house may need either turbine fans powered by the wind or electrically operated power vents. Either way, adequate ventilation of the attic is a must to reduce cooling costs and moisture in the attic and to help the insulation system in the home do its job properly.

The choices made in framing, roofing, and siding affect subsequent steps in the construction process. Special care must be given to planning the framing process, the roofing material, the doors and windows, and the type of siding—whether brick, stone, stucco, vinyl, aluminum, steel, or wood—because each of these choices affects the final result. After the steps for framing, roofing, and siding are completed, the builder can move on to the plumbing, heating, and electrical work for the house.

4

Plumbing, Heating, and Electrical Work

After the framing and often while installing the siding, the builder begins to rough in or place the plumbing, heating, and electrical systems in the walls of the house behind the drywall or interior paneling. In many new homes, the builder also installs a central cooling system.

In addition, some common low-voltage components of the home—such as telephones, cable televisions, burglar alarms, intercoms, doorbells, thermostats, and smoke alarms—are usually prewired at this point. Many modern homes may also have a central vacuum or other specialized features that require some prewiring before the builder installs the drywall or paneling. Today, the builder can even integrate these electrical-power and signal-wiring systems into a SMARTHOUSE wiring system controlled by a central computer.

Once the builder has had the subcontractors rough in the plumbing and electrical systems and before the drywall or paneling goes up, the local municipal inspectors examine the roughed-in portion of these systems for compliance with local codes.

Plumbing

At the plumbing rough-in stage, the builder needs to know the exact location of all plumbing fixtures because the plumber will now install the plumbing lines for both the hot and cold water supply and the sewage. (See photo 4.1.) To install the plumbing lines, the plumber usually drills large holes through joists, studs, bottom plates, and subflooring. Local building codes detail specifications for the size, location, and number of these holes. The actual plumbing fixtures will be installed later as the house nears completion.

Usually the plumber also installs the tubs and shower bases at this point. In many areas, inspectors require the plumber to seal the entire plumbing system and pressurize the supply pipes to check for leaks. If any leaks appear at the rough-in inspection, the builder must then call the plumbers back to fix them.

In some areas, after installing the plumbing lines, the builder may fill the holes with some type of hardening foam. This increases the energy efficiency of the exterior walls. It also helps to keep out pests, such as mice and roaches, and

4.1 Plumbing Rough-in

prevent the spread of fire from one floor to another. The builder may install this foam at the same time as the insulation, after putting up all the mechanical systems. In some municipalities, however, inspectors require the builder to install the foam separately before the insulation, so the inspector can easily verify that the builder has properly filled all holes.

A few other points about the plumbing may interest you. You may notice that portions of the sewage pipes extend up through the roof. These are the vent pipes discussed in Chapter 3. The sewage vents balance air pressure on both sides of traps, which prevent sewer gas from entering the house through sink drains, and allow the gas to vent to the outside. Collars installed around these vent pipes at the roof prevent water leakage.

As part of the plumbing system, the builder may also place one or more hose bibs outside the home for watering the lawn or washing the car. The location of these bibs can vary, so the builder and homeowner will want to work together to select these locations.

Heating and Cooling Systems

The central heating and cooling ductwork is generally installed after the plumbing. (See photo 4.2.) The installation of these two systems takes so much work by so many people that the builder may avoid scheduling them at the same time. The choice of which step comes first is usually the builder's and subcontractor's preference.

A house can have a room heating or cooling system with a separate unit and thermostat in each room. For example, an electric-resistance baseboard heating system has individual heaters and thermostats in each room. As an alternative, a house can have a central heating and cooling system with one unit, thermostat, and distribution system to carry heat and cool air throughout the house. The most common types of central heating and cooling systems are a natural gas furnace with an electric air conditioner or an electric heat pump that provides both heating and cooling.

4.2 Ductwork

Natural Gas

The typical natural gas heating and cooling system has a gas furnace with a separate electric cooling compressor mounted outside of the house. The system also comes as a packaged unit that the builder can mount on the roof or other outside locations. Functionally the two types of gas systems are identical.

In the summer, the system pulls warm air from within the house into return air ducts and passes it across a set of cooling coils. The cooling coils contain a gas similar to the gas in your refrigerator. This gas absorbs the hot air in the house and cycles it to a compressor located outdoors. The cooled air is then distributed throughout the house through the supply ducts.

In the winter, the air handler pulls cool air from within the house and passes it through a gas heating unit. The warm air is then recirculated through the same ductwork used for cooling in the summer.

Like your refrigerator, an electric cooling system is a closed system requiring little or no maintenance. Although it may need recharging at some point, if properly maintained, this system should give years of trouble-free service.

The gas furnace and air conditioning system share a common blower, a common thermostat system, and common ductwork, so they seem like one system when they are really two distinct systems. If a house has this kind of system, the builder has a certified gas plumber run gas lines, which look like plumbing lines, to the furnace.

Electric Heat Pump

A heat pump is essentially a reversible air conditioner. The heat pump may look and sound like a stand-alone compressor or packaged unit. While the gas furnace blows hot air in the winter and the air conditioner blows cool air in the summer, the heat pump, which operates on electricity, is a single system designed to both heat and cool.

Heat pumps follow the principle that outdoor air contains heat or thermal energy even in the winter. During the winter months, the heat pump extracts heat from the outdoor air and circulates that heat through the house. In the summer, the system is reversed. The heat pump removes heat from indoor air, discharges the heat outdoors, and circulates the cooled air through the house.

Heat pumps are efficient systems in moderate climates if the home is properly insulated. Additionally, the heat pump contains back-up or emergency electric heating coils that automatically turn on if the outside temperature falls below about 25 to 30 degrees.

Ceiling fans can also be an integral part of a home's cooling system, particularly for warm spring and fall evenings. If the extensive use of ceiling fans is planned for a new home, the builder will have the electrician prewire switch controls for the fan and optional light kit.

Electrical Wiring

Since installation of the plumbing and ductwork can require large cuts in the wall and floor systems, builders usually leave the rough-in of the electrical wiring until last. In most jurisdictions, the utility company is responsible for setting the meter and installing the line to the transformer and electricians are responsible for running electrical lines from the meter into the home. In some areas, however, electricians may also furnish the meter. (See photo 4.3.)

*4.3 Electrical Meter Box
Ready for Meter Installation*

As the first step, the electrician installs a service wire that leads from the meter into the breaker box. The breaker box distributes electricity by circuits or wires to outlets in each room of the house, as well as the thermostat, light switches, the doorbell, appliances, the intercom, and the security system. Most electricians try to locate the meter and breaker box as close as possible to one another in a convenient location in the house.

The electrician next sizes the breaker box to handle the electrical load expected for the home. Most single-family detached homes with no special requirements call for a breaker box with at least a 200-amp capacity. However, if the homeowner plans to add on to the home or run any special circuits later, the service should be upgraded accordingly.

Upgrading the electrical system at this time is not nearly as costly as upgrading the service at a later date.

From the breaker box, the electrician then runs wires for all of the circuits in the home. The electrician usually threads these wires through holes drilled in the stud walls and joists or staples wires to the framing in inaccessible areas.

You may notice that the wires for different circuits come in different sizes and types. Most outlets have a 120-volt service. In general, the higher the amperage expected for a given circuit, the larger is the wire needed to conduct that amperage. Also, certain appliances in the home will require a 240-volt service. These appliances may include the hot water heater, dryer, range, heat pump, and air conditioner.

Breakers of various ratings protect all of the circuits in a home and are designed to trip when they detect an overloaded circuit. This prevents the wire or the appliance from getting hot and starting a fire. However, the breakers also allow a brief period of overload before tripping, since most appliances, particularly those with motors, run for a short time at a high amperage before settling down to their normal levels.

Circuits that carry currents to bathrooms and outdoor outlets or appliances need more sensitive breakers to reduce the danger of electrical shock in areas where plumbing and water are commonly found. For those circuits, the electrician uses ground-fault circuit-interrupters, which trip the moment a faulty appliance such as a hairdryer is grounded.

The notion of an electrical circuit may seem a little confusing. Some appliances like refrigerators, dryers, or ranges are such a drain on the electricity that they need their own circuits. On the other hand, circuits for bedrooms can typically handle several receptacles and lights. If the house has high-wattage circuits for such things as overhead fans, waterbed heaters, or window air conditioners, the electrician may need to decrease the number of outlets on a circuit accordingly. However, with normal loads, just a few circuits can handle the low-usage areas of the home.

As a final step in the electrical wiring, the electrician mounts the boxes for the outlets, switches, and fixtures before the builder moves on to the insulation of the house.

Insulation

After the builder has had the plumbing, heating, and electrical components roughed in but before the drywall or paneling goes up, the builder installs the wall insulation.

4.4 *Insulation of Walls and Vaulted Ceiling*

Eventually, the floors and ceilings also need insulation. (See photo 4.4.) However, if access to the floors and ceiling is available through a crawl space and attic, then the builder can insulate these areas later.

Wall Insulation. The most common wall insulation is fiberglass or rock wool (mineral wool) made of very fine fibers spun from molten rock or glass. This type of insulation may come either as faced or unfaced flexible blankets or as rigid or semi-rigid boards. Finally, fiberglass or rock wool may come in a loose fiber form, which the insulation subcontractor can blow or pour into cavities like attic spaces.

Blankets used for insulation generally have a kraft paper or aluminum foil facing that may serve as a vapor retarder to prevent moisture from collecting in the walls. The blankets are the width and length of the cavity between the studs and the paper extends an inch or so on either side of the studs to serve as a stapling edge.

If the builder decides to install unfaced fiberglass insulation in the stud walls, the inner face of the wall should be covered with a material such as polyethylene to serve as a vapor retarder. Vapor retarders must always face the conditioned space—in on walls, down on ceilings, and up on floors. However, a vapor retarder is not always necessary on the ceiling if the attic space is well ventilated.

Attic Insulation. The builder normally installs the attic insulation after the gypsum board is installed on the ceiling. On flat attic spaces, the insulation subcontractor uses either fiberglass blankets or a blown insulation made of fiberglass, mineral wool, or cellulose. Either approach provides effective insulation for attic areas.

Blown fiber insulation has the advantage that it can fill nooks and crannies of the attic that the insulation subcontractor cannot reach with blankets. However, blown insulation requires heavy equipment and can also be messy to install, which explains why many builders try to complete this phase soon after the gypsum board is installed in the ceiling.

Ceiling Insulation. Spaces above vaulted ceilings require fiberglass blankets as insulation, which the builder needs to install before the drywall. (See photo 4.4.) The ceiling insulation provides a blanket to keep the home warm in winter and cool in summer. Vaulted ceilings present one additional challenge for the insulation subcontractor. Ventilation of the space between the insulation and the roof sheathing is every bit as important as the insulation of that space.

Proper ventilation for the insulation of ceilings, as well as attics, provides the following for a house:

- allows moisture to escape
- maintains the R-value of the insulation (The R-value indicates the level of resistance to heat flow in a building

material. The higher a material's R-value, the more effective insulation it provides.)
- prevents condensation from collecting within the insulation or on the bottom of the sheathing
- prevents mold and rot
- cools the roof in the summer

In colder climates, proper ventilation of roof cavities also keeps the roof cold in the winter and thus reduces the melting of snow on the roof and ice damming at the eaves.

The insulation subcontractor can combat the buildup of moisture in ceiling spaces by installing insulation baffles that provide an air channel from the eave vent, through the space between the rafters. The insulation subcontractor nails these baffles to the underside of the roof sheathing before installing the insulation.

With this type of insulation, you can see that the continuous eave and ridge vent system is virtually a must in homes with vaulted ceilings. On the other hand, homes with flat ceilings can use eave vents, gable louvers, ridge vents, or other types of vents to provide ventilation.

Floor Insulation. The builder installs the floor insulation between the floor joists by using short pieces of stiff wire, sometimes called dogs or tiger teeth, which press the batts up against the subflooring. In heated basements, the walls are usually insulated instead of the floor.

In crawl spaces in very cold climates, the builder may also install some form of foundation wall insulation. This typically involves plastic foam insulation boards or fiberglass blankets applied to the inside of the foundation wall within the crawl space. For spot insulation the builder may have a plastic foam sprayed into the cracks around windows and doors, pipe openings, and other air leakage points. The builder may also have a plastic sheeting spread on the ground within the crawl space to control moisture.

All of this begs the question, "How much insulation is enough?" First, of course, local climate conditions determine the amount of insulation a house requires. In addition, in some areas, local building codes require minimum amounts of insulation in a house. Generally, insulation is inexpensive for the job it does, and it usually pays to go for the most insulation possible. However, remember that energy conservation is more than just pouring on the insulation.

Multipane windows with low-E glass, careful caulking and weatherstripping of doors and windows, and use of south-facing windows and high-efficiency heating and air conditioning equipment are also important. The home builder and public utility company can provide sound advice on the level of insulation required for a house in a particular climate.

The house is now a maze of plumbing pipes, wires winding through studs, ducts for heating and cooling, and insulation of all kinds. As a final step, the builder now roughs in the specialty components of the house.

Miscellaneous Rough-in

Before the drywall goes up, the builder roughs in such common specialty components as the wires for telephones, doorbells, intercoms, security systems, and cable televisions. Again, as another alternative, SMARTHOUSE wiring integrates all of these wires into one system.

Virtually all homes have telephones, and having the electrician prewire the appropriate rooms at the rough-in stage is an inexpensive way to ensure that the homeowner has phone service wherever required. Wiring two phone lines at the same time is usually no more expensive than wiring one. In this way, the homeowner will have access to wiring for that home office or teenager who wants a phone later on.

Cable television is a different challenge. Prewiring for cable ensures that unsightly cables do not run along the baseboard or through holes drilled in the carpet. However, prewiring cable service to every room of the house is not as feasible as prewiring for phone service. With both telephone and cable service, the homeowner should specify exactly where the outlets should go to match furniture placement and lifestyle choices.

Many homes today are also fitted for burglar alarms. These come in a variety of styles and models, and the builder can recommend a suitable alarm company. Prewiring for a basic alarm system is much less expensive than wiring for the system after the home is finished, even if the homeowner plans to have the system installed later. Many alarm companies prewire on a cost basis and anticipate that homeowners will purchase the system soon after closing.

Other specialty items that the builder may rough in at this point include such amenities as a central vacuum (see photo 4.5), intercom, smoke alarm, built-in multiroom stereo system, computerized climate control, and controls for a swimming pool. Also, an automatic sprinkler system controlled

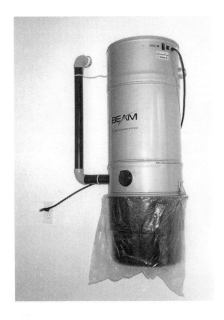

4.5 Central Vacuum

from the garage, certain types of ceiling lighting, icemakers, a sauna, and a whirlpool tub can be added to the list of amenities that a builder may rough in at this stage in even moderately priced homes.

All of these amenities are listed to make a point. If the homeowner plans to add any of these features after the home is completed, the builder should really do the preliminary rough-in for these features at this stage of construction. Although the builder cannot dig the hole for the swimming pool planned for next year, completing the prewiring or pre-plumbing for this type of additional amenity is much less expensive at this phase.

So, while the process may vary somewhat, to rough in the plumbing, heating, and electrical systems of the house, the builder first installs the plumbing and then the heating and cooling systems. After completing these steps, the builder next puts in the electrical wiring and begins to place the insulation in the walls, attic, ceilings, and floors. At this point any specialty components for the house are also roughed in before the builder moves on to the interior finish.

5

Interior Finish

Although you may find the steps described in this chapter easier to visualize, just remember that the drywall, trim, and cabinets only turn out right if the foundation, walls, and rough structure have been carefully planned and constructed. Building a house is like putting together a complex puzzle. All the pieces are interconnected, and the puzzle is complete only when all the pieces are in place.

Fireplace and Other Interior Brickwork

Where used, one of the first pieces of the interior puzzle to be installed is the fireplace. The builder may complete the fireplace soon after roughing in the plumbing, heating, and electrical systems, although some builders may wait until later. While a fireplace is not essential, it is a traditional part of many American homes.

The most traditional type of fireplace is the brick fireplace pictured in drawing 5.1. The builder can also choose many alternatives to the brick fireplace, including metal prefabricated fireplaces or wood stoves. In addition, the builder can build traditional fireplaces out of field stone, cut stone, or other masonry materials besides brick. However, regardless of the type of masonry used for the facade, the builder usually constructs the interior, functioning part of the fireplace as shown in drawing 5.1.

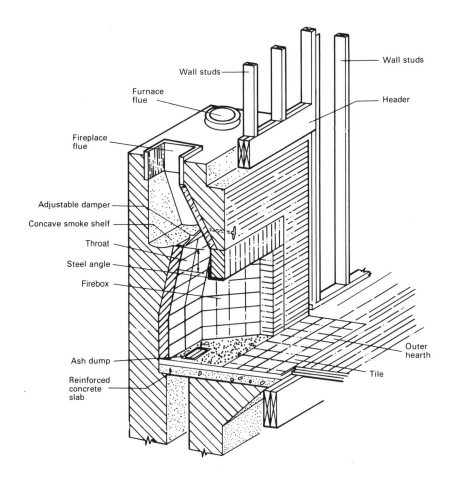

5.1 Components of a Masonry Fireplace

Traditional fireplaces at best are only marginally efficient in heat production. For that reason, they are more of a luxury than a source of heat for the home. In drawing 5.1, you can see the firebox where the actual fire is built and the flue, which rises from it. A steel angle, also called an angle iron, runs from one side of the firebox to the other and supports the brick facade over the opening.

A metal damper separates the smoke shelf from the firebox. The homeowner must open this damper before building a fire to prevent the room from filling with smoke. When the homeowner is not using the fireplace, the damper should be closed to keep interior air, usually expensively heated or cooled, from escaping outside.

Depending on the overall size of the firebox, the back of the fireplace is usually narrower than the front and slopes toward the front at the top. This helps channel the smoke and fumes toward the rear of the firebox and ultimately into the flue. Likewise, the smoke shelf at the bottom of the flue helps to direct any downdraft back up the flue and keep smoke and fumes from reentering the firebox. All in all, a masonry fireplace is a complicated structure and requires the talents of trained masons experienced in fireplace construction.

The builder can increase the heating capacity of a fireplace by using an air-circulating fireplace insert with double steel walls. A common type of insert is a fireplace form, which allows cool air from the room to enter the system at the bottom between the double walls, be warmed by the inner steel wall which is in contact with the fire, circulate upward, and reenter the room through upper vents.

In addition, the builder can add small fans to circulate the air and further increase the efficiency of the fireplace. The builder can also have outside air flow through ducts into the firebox for combustion and can add glass doors to the front of the firebox to reduce the amount of heated room air that goes up the chimney.

Many builders have found these types of fireplace systems helpful, particularly in heating family rooms or dens. Of course, the usefulness of any fireplace system also depends on the local climate, the cost of firewood, and the effort that the homeowner is willing to exert in cutting and splitting wood for fuel.

Finally, besides the interior brickwork needed for fireplaces, the builder may use brickwork in many other places in the house, including decorative interior brick walls, solid bricks for floors, and even brickwork for interior staircases. Glass block, commonly used many years ago, is making a comeback in bathroom design and other decorative features. A builder usually hires a trained mason to properly install these types of masonry in a house.

Plaster and Drywall

Two important pieces of the puzzle that demand the builder's attention early on in the interior finish are the plaster and drywall.

Plaster

A generation ago builders commonly used plaster in most homes. Today gypsum wallboard, commonly called drywall, has almost completely replaced plaster. However, plaster still has certain applications, particularly in remodeling and renovating existing plaster homes or in creating designs for molded decorative trim.

However, for the most part, old-fashioned, site-cast plaster moldings have now given way to factory-cast plaster and plastic foam moldings. These days chances are good that a builder can find a ready-made substitute for plaster decorative features from older homes. These substitutes are less expensive and have the same look and feel as plaster.

Drywall

Drywall is basically a board that the builder applies to the frame of the house to create the walls. It typically comes in 4x8- and 4x12-foot pieces. Professional drywall installers often prefer 4x12-foot pieces since they require less cutting and patching.

Drywall for Ceilings. In square or rectangular rooms with standard, flat ceilings, the builder usually installs the ceiling drywall first. To start the process, the professional drywall installers carefully measure and cut the drywall on the floor to make sure that each end of the drywall falls exactly halfway over a ceiling joist or truss.

The drywall installers next lift a piece of drywall into place in one corner of the room, with the long dimension of the drywall running perpendicular to the ceiling joists or trusses. They then nail or screw this piece of drywall into each joist or truss. Drywall screws cost somewhat more than drywall nails but are less likely to push out from the wood as it dries, causing a bump in the surface called a nail pop. Adhesive may also be used to reduce the number of nails or screws needed for installing the drywall.

Drywall for Walls. As the first step to placing drywall on the walls, the installers run the drywall lengthwise along the wall. A house usually has walls that are just over 8 feet tall. The height of the walls includes the height of the studs and bottom and top plates.

Since the installers usually prefer to use drywall that comes in 4x12-foot pieces, they next cover the wall with two pieces of drywall, an upper piece along the ceiling and a lower piece along the floor. They then install the upper piece first to avoid any gaps at the ceiling. They next install the lower piece to butt up against the upper piece. This leaves a small gap where the drywall meets the floor, which will later be covered with carpet or baseboard.

The drywall installers try to stagger upper and lower ends of the sheets every 4 feet. This approach strengthens the wall and makes the vertical joints less visible after painting. Except at corners, the ends of the upper and lower sheets are not placed on the same stud.

You may notice that in some places the builder uses specialty drywall. For example, thicker drywall or other special fire-resistant gypsum board is sometimes used between the house and garage. Also, while white or gray paper covers most drywall, the builder places a special water-resistant drywall in high moisture areas like bathrooms and laundry rooms. This drywall is covered in green paper and derives its water resistance from a wax emulsion.

Sometimes the builder may use drywall panels with factory-installed decorative vinyl wall coverings and pre-finished edges. To install these panels, the builder glues, rather than nails or screws, them to the studs to keep from destroying the decorative pattern.

Otherwise, to finish installing the drywall, the installers drive the heads of the drywall nails or screws slightly below the surface, which leaves a slight dimple in the surface of the drywall. The drywall finishers then apply a drywall finishing or joint compound to fill these dimples. This compound comes as both a powder or, more commonly, premixed in

5.2 Drywall Installed with Tape and Joint Compound

5-gallon buckets, which is what most drywall finishers prefer to use.

Besides using this drywall finishing compound to fill the dimples in the surface of the drywall, the finishers also use it to apply a layer of 2-inch-wide drywall tape, which is simply heavy paper, over the edges of the drywall joints to smooth the surface and reduce the chances of cracking. (See photo 5.2.) The finishers then use a wide spackling knife, somewhat like a putty knife, to spread the compound onto the edges of the drywall joints. With the knife, they force the compound into the recesses of the joints and dimples. They then apply the compound to the tape and joints.

The joints and dimples in the drywall must then be allowed to dry between coats, usually overnight, although the finishing compound can take a lot longer to dry in damp or cold weather. Because the compound shrinks below the surface as it dries, the drywall finishers then apply a second coat of compound with a wider knife. After the joints and dimples in the drywall are dry and smooth, the finishers sand them to prepare the drywall for painting, wallpapering, or the application of textured coatings.

Exposed exterior corners of the drywall are particularly vulnerable to damage. When the drywall is installed, the builder covers these corners with an angled metal corner piece, which comes in 8-foot lengths. The purpose of this metal corner piece, sometimes called a corner bead, is to provide rigidity and strength to the corners of the drywall. The finishers cover this metal corner piece with two or three coats of compound so that it is indistinguishable from the drywall that it protects.

When installing and finishing the drywall, a builder's methods may vary, whether for marking and cutting holes for electrical boxes, dealing with difficult corners, or finishing smooth or textured ceilings or walls. Regardless of the techniques used, however, all of these methods require experienced, professional drywall specialists to complete this very basic part of the finish for the home. A professional home builder will have one or more experienced drywall crews, either working directly for the builder or more commonly subcontracted, who can provide a proper drywalling job for the house. After finishing the installation of the drywall, the builder can begin to put the interior trim in place.

Interior Trim

Installing the interior trim has four basic steps: door and window trim, moldings, cabinets, and finish hardware. In addition, the interior staircases are usually completed at this point. The steps to interior trim might not happen at the same time nor may the same carpenters complete each step.

For example, a builder often subcontracts cabinet installation, particularly if the cabinets are built off site or provided by a specialty cabinet company. Although this approach is increasingly rare today, if the cabinets are built on site, then the builder may hire a custom cabinet craftsperson. This person and any assistants need undisturbed access to the house until they finish installing the cabinets.

In addition, the builder usually has an interior trim crew, either on the payroll or subcontracted, who installs the interior doors, door and window trim, baseboard, and other moldings. Finally, after the staircases and other interior finish work are completed, a carpenter installs the finish hardware.

Door and Window Trim

The builder needs to select moldings for the doors and windows early on. This is because the doors, hinges, and trim are usually ordered from a factory in advance where they are preassembled and then delivered to the home site at the proper time.

The builder or homebuyer can select finish trim from a variety of moldings for around doors and windows and the intersection of the walls with the floor and ceiling. Moldings that can be finished naturally are often made of oak or other hardwood, while moldings that the builder plans to paint are usually made of softer wood. Prefinished molded wood fiber and plastic have also recently met with success in some applications for moldings.

Doors and Door Trim. House plans show doors partially open, so the builder and the homebuyer can tell which way the door is supposed to swing and whether any obstructions may prevent the door from opening or closing in that direction. Bedroom and bathroom doors usually swing into those rooms because if these doors swing into hallways they may interfere with traffic. Closet doors usually swing outward, because closets have no room for doors to swing inward. Doors between two rooms, such as a kitchen and dining room, swing whichever direction will least interfere with future furniture placement and traffic patterns.

Interior doors usually come approximately 6 feet and 8 inches high and in a variety of widths. The most common width for bedroom doors is 2 feet and 6 inches, while bathroom doors often range from 2 feet to 2 feet and 6 inches wide. The widths of closet doors are in the same range, although walk-in closets often have wider doors. If the homeowner anticipates needing wider doors for wheelchair access, he or she may want to work with the builder to plan for wide-passage doors—at least on floors where a wheelchair may be used.

Closet doors come in a variety of types and styles. For example, closets may have bi-fold doors, which are really two doors hinged together. Each pair of doors is then hinged at the side so that the doors fold open and close accordion style on a track at the top and sometimes the bottom of the door opening. Bi-fold doors come in patterns to match the other doors of the house or in other patterns such as louvered styles that are particularly useful for utility closets.

The builder may also use special accordion or sliding doors for closets. An accordion door has a series of hinged vertical components, made of material like vinyl, that typically fold open and close on rollers along an overhead track like an accordion. Sliding doors also move on rollers in a track at the top or bottom of the door and usually slide in front of each other in the door opening. For some closets and other places in the house where neither swing direction is appropriate, the builder may use a sliding pocket door. For convenience this type of door slides on a track into a cavity or pocket in the wall.

Bi-fold, accordion, sliding, and pocket doors are usually not prehung but rather are delivered with a hardware kit. A skilled trim carpenter then assembles and installs these doors on site. The builder needs to plan all doors and door openings at the framing stage, because the rough openings for doors can vary greatly in size and thus affect the wall framing. As one example, the pocket door frame has to be planned as an integral part of the adjacent wall from the beginning.

As the final step to installing the interior doors, the builder applies the casing trim around the door openings. As part of this process, the trim carpenter nails the door casings to the door jamb or door frame and through the drywall into the wall framing with finish nails. To secure the hardwood casings, the carpenter may have to insert nails in moldings through predrilled holes, which prevent them from splitting.

While trimming the interior doors, the builder also trims the interior windows.

Window Trim. The casings around window frames on the interior of the house should match those used around the interior door frames. Since molding patterns may vary among suppliers, the builder needs to coordinate the style of the casings used to finish the windows with the style of the pre-hung doors to ensure that the desired moldings are supplied.

Besides the casing, interior trim for a window includes the stool and apron. (See drawing 5.3.) The stool is the ledge located at the bottom of the window on the interior of the house. The apron is a piece of molding similar to the casing used below the stool. Window stools are sometimes pre-manufactured with an angled notch to match the angle and depth of the window sill.

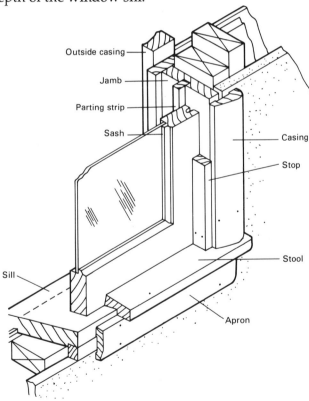

Outside casing

Jamb

Parting strip

Sash

Casing

Stop

Stool

Sill

Apron

5.3 *Window Trim*

Windows also have several other parts. The sash is that part of the window that contains the glass. The sash stop is a narrow vertical strip that holds the sash in place. This stop usually comes as part of the window unit, but a molded stop similar in style to the casing may also be used with traditional wood windows.

Window casings, stools, and aprons come in many sizes and patterns and certain types may match a particular style of window better than others. With prebuilt window frames, the other trim shown in drawing 5.3 usually comes preinstalled. For example, the parting strip, which keeps the upper and lower sashes of a double-hung window apart, is often part of a premolded plastic jamb liner. This liner also contains a spring device that aids in opening the sashes and keeps them from falling down when they are in the up position.

Interior Moldings

As the builder completes the installation of the interior door and window trim, the builder also puts other interior moldings in place. Interior moldings include base, ceiling, and wall moldings.

Base Moldings. Sometimes called base or baseboard, base moldings serve to conceal the joint between the finished wall and the floor. They come in several widths and styles and should generally match the style of the interior door and window casings, such as contemporary clamshell or traditional colonial. In custom work, the builder may use a two-piece base molding, consisting of a large baseboard capped with a small piece of decorative molding called a base cap. This smaller molding serves to conform closely with any variations in the wall.

When the finish floor material is hardwood or vinyl, then the builder may attach a base shoe to the base molding. The base shoe is a strip of molding that covers the intersection of the floor and base molding and conforms with any variations in the floor. When using stained hardwood for floor covering, the builder may install a hardwood baseboard that is stained to match the wood floor.

If the floor covering is carpet, then the builder usually eliminates the base shoe. As an alternative, when installing vinyl flooring in bathrooms and kitchens, the builder often uses a rubber base molding because it will not be damaged when the floor is washed. Similarly, ceramic base pieces are often installed with ceramic tile flooring.

Ceiling and Wall Moldings. Particularly in traditional homes, the builder sometimes uses crown moldings at the junction of walls and ceilings for architectural effect. In contemporary or vaulted ceilings, however, these moldings are typically not used. Also, in dining rooms of traditional homes, the builder may install a chair rail parallel to the floor at about the height of the back of a dining room chair. This molding serves both a functional and decorative purpose because it protects the wall from damage when the chairs are pushed against the wall.

Finally, the builder may want to apply moldings and trim to other parts of the home, particularly in more traditional styles of homes where the wood craft techniques of past generations might be especially appropriate. With modern materials and tools, the builder can readily duplicate many of the details used in older homes.

Cabinets and Other Millwork

As a general term, millwork includes most of those house components manufactured from wood or wood products, including door and window frames, moldings, stairs, and cabinets. While some traditional cabinets are built on site, this approach has become rare. Instead, today most builders purchase cabinets—whether for the kitchen, bathroom, laundry, or other parts of the house—from a specialty cabinet manufacturer.

Kitchen Cabinets. The kitchen usually contains more millwork than the rest of the house. Manufacturers construct kitchen cabinets, both base and wall units, to a standard height and depth. For example, the normal height for the base cabinets is 36 inches and appliances for the kitchen, such as dishwashers and ranges, are manufactured with this in mind. (See photo 5.4.)

5.4 Installation of Kitchen Cabinets

Wall cabinets vary in height, depending on their location. Typical wall cabinets are 30 inches high. Traditionally wall cabinets do not extend to the ceiling. Furthermore, the top of wall cabinets is usually set at 7 feet because at that height the area is inaccessible to the average person.

Today, the modern kitchen has become a storehouse for more stuff than is found in a typical Army mess hall. As two-wage-earner families have increasingly turned the kitchen into the center of activity, the number of kitchen appliances, implements, and specialty food products has increased dramatically. Some millwork manufacturers have responded to this trend with special ceiling-height cabinets to increase the storage area in the kitchen.

The array of specialized kitchen cabinets seems endless. For example, the builder can place cabinets over a sink or stove. The cabinet over the stove is usually 18 inches high to provide space for a vent hood with a light, either externally vented or internally filtered. In addition, the builder usually installs a 12-inch-high cabinet over the refrigerator.

A kitchen may have a variety of other storage cabinets, such as a lazy-Susan cabinet in an otherwise unusable corner, which makes for efficient use of space. Some kitchens may also have special cabinets with sliding racks to aid in storing big pots and pans or rotating recycling bins. Other kitchens may have special cabinets used to mount microwave ovens or, for that matter, to mount built-in ovens, convection ovens, or any combination of these in a handy arrangement.

Just as the selection and coordination of kitchen cabinets must be carefully planned, the builder must also plan and coordinate the installation of the bathroom cabinets.

Bathroom Cabinets. Builders often purchase bathroom cabinets from the same millwork manufacturer that provides the kitchen cabinets. Like kitchen cabinets, manufacturers build bathroom cabinets to preestablished standard sizes, which ensures that the sinks and vanity tops will match. This is critical, for example, if the builder wants to use a molded plastic sink that is an integral part of a plastic countertop in the bathroom.

While natural wood cabinets are available for bathrooms, builders often use plastic laminates because of the heavy exposure to moisture in the bathroom. Countertops come with single or double sinks and the cabinets must conform to the number and location of the sinks. The choice of bathroom cabinets and sinks must be made early, since it affects how the builder roughs in the plumbing for the bath.

Bedroom Closets. Since modern working Americans spend an increasing amount of time in the bedroom and bath, these days even the most modest homes have a great deal of design detail placed on storage space in the master bedroom. Most homes have significantly better designed and larger closets in the master bedroom than the average home of a generation ago. The need for storage space has also extended to the other bedrooms of the house.

As bedroom closets get larger, the desire for more efficient shelves and rods may also increase. Long, straight pieces of wood shelving are increasingly harder to find and relatively expensive. As a result, many builders now use particleboard shelves or premanufactured wire shelving, which is more adaptable to a variety of storage requirements. While the installation of this type of shelving takes less time, skill, and tools than working with shelving board, it still requires a trained and experienced person to install.

Other Millwork. A typical home may also have other pre-manufactured millwork items. For example, the fireplace may need some type of decorative mantel. With a colonial or traditional interior styling, a well-designed fireplace mantel may have both a shelf over the fireplace and decorative wood trim beneath the shelf. The builder must locate the mantel at least 12½ inches above the opening of the fireplace.

The formal rooms of some houses may even have elaborate wood cabinets that surround the fireplace. In these situations, the builder ensures that no wood or other combustible trim material is placed within a certain distance from the edge of the fireplace opening.

As the builder installs the door and window trim, interior moldings, cabinets, and other millwork, the builder also usually completes the construction of the interior staircases. However, the preliminary stages of staircase construction really begin earlier in the process.

Staircase Construction

The construction of an interior staircase begins at the framing stage. On the plans the designer locates the staircase between the first and second floor and, if relevant, between the first floor and the basement of the home with a careful eye toward anticipated traffic patterns and aesthetics.

As the first step in installing a traditional staircase, the framing carpenters put up the actual framing for the steps called the stringer. (See drawing 5.5.) They then cover the stringer with temporary treads, usually scrap 2x4s or 2x6s, for the workers' use. Later, during the interior trim phase, the trim carpenters remove the temporary treads and install the finish treads. The finish treads may be made of oak or other hardwood, if the builder plans to stain them. Other-

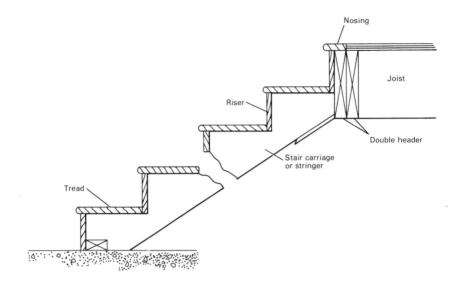

5.5 Staircase Assembly

wise, the builder may finish both the treads and the risers with finish lumber and cover them with carpet.

If two walls completely contain the staircase, the trim carpenters usually attach a hand rail to one wall of the stairwell. On the other hand, if the stairwell is exposed on one or both sides, then the trim carpenters install decorative guard rails or balustrades at the open sides of the stairs.

A few other important points about staircase construction are worth noting. First, the builder must make the finished staircase a minimum of 36 inches wide. That width provides sufficient room for moving furniture and, more importantly, people between floors. Also, most building codes require a minimum headroom clearance in a stairwell of 6 feet and 8 inches at any point on the stairs. The stairwell may be left entirely open to the second story, or the builder can put a sloped ceiling above the stairs.

The builder calculates the width and length of the opening for the stairwell before framing begins, so the framing crew knows the exact location and dimensions of the staircase. Of course, these measurements as described assume a straight stairwell. However, many staircases turn at least one right angle and require a landing or may even have a curve. In those cases, the lead carpenter on the framing crew consults with the builder on the site to assure proper placement of the stairwell.

Today most new, two-story homes have staircases built in a millwork shop rather than on site. With this approach, the millwork shop prebuilds the staircases and then delivers them as packages for easy installation. Of course, the builder must still assure that these prebuilt units are properly installed in the home.

After installing the door and window trim, moldings, cabinets, stairs, and other millwork, the builder moves on to the final step of the interior trim—the installation of the finish hardware.

Finish Hardware

The builder may wait to install the finish hardware, since painting and other finish work can damage these items. However, the interior trim is not really complete until the finish hardware is in place. Finish hardware appears in many areas of the house, especially on the doors and windows and in the kitchen and bathrooms.

Door and Window Hardware. The most common types of door and window hardware are doorknobs, doorstops, and window locks. This hardware comes in a variety of finishes. The two most common are antique brass and polished brass. Builders may also occasionally use chrome hardware for these items. Since the manufacturer installs the hinges on the doors at the millwork factory, the builder selects this trim item early and coordinates it with the other door and window hardware for the house.

Interior doorknobs come in two general types: privacy and passage knobs. Privacy knobs are lockable from one side, although they usually have some feature that allows the homeowner to open the door from the other side in emergencies. Builders often use privacy knobs for bedrooms and baths. Unlike privacy knobs, passage knobs are not lockable. Builders may use these for closets and other rooms that do not need to be locked. Most homebuyers want exterior door locks and dead bolts to match the interior knobs. To ensure that all these items match, the builder usually buys all of the locks from the same supplier.

A doorstop is a piece of hardware that prevents the door and doorknob from slamming into the wall. The old-fashioned kind of doorstop is a decorative brass prop with a screw tip on one end and a rubber bumper on the other. To install this type of doorstop, the builder simply screws it into the baseboard. Once installed, the door hits the rubber bumper before it hits the adjacent wall.

However, in many homes today, the builder has replaced the traditional rigid prop with a spring that does not break off so easily if it is accidentally kicked. Also, another type of doorstop has recently been invented called the hinge-pin doorstop. Rather than screwing into the baseboard, this doorstop attaches to the pin in one of the hinges. It is adjustable to prevent the door from going all the way back against the adjacent wall.

Window locks come in many varieties, but the most common is hardly different from the latches that builders have used for several generations. This type of latch is popular because it is easy to open, which is handy if the homeowner plans to use the windows for ventilation. The drawback to this type of latch, however, is that it may be less secure. If the homeowner does not plan to use the windows for ventilation, then the builder may want to install more secure locks, such as barrel locks. Today many manufacturers offer such locks with the windows, so the builder does not necessarily have to install them separately.

Kitchen Hardware. The kitchen hardware consists of cabinet doorknobs, hinges, and drawer pulls. Some cabinets also come with recessed finger grips instead of knobs. When installing kitchen hardware, the builder will generally coordinate the style of the hardware with the style and color of the cabinets, light fixtures, appliances, and the overall color scheme of the kitchen.

Bathroom Hardware. Hardware for the bathroom includes such items as towel bars, tissue holders, toothbrush holders, and soap dishes. The homeowner will probably want all of this hardware to match each other and the knobs on the faucets and toilets provided by the plumber. So the builder must coordinate the selection of bathroom hardware to ensure that the bathroom turns out aesthetically pleasing.

Other Finish Hardware. Many other hardware items may be located throughout a home. One example is a brass kick plate for the front door. This is a plate placed at the bottom of the front door to protect it from marring. Other hardware items in the house may include accent pieces for the dining room and decorative light switch and receptacle plates.

For starter homes, the builder usually specifies the hardware to keep the costs down. For larger, move-up homes, the builder may provide a hardware allowance. The buyer must then carefully select the hardware, perhaps with the guidance of the builder's recommended hardware suppliers.

Before the actual finish of the home but usually after all of the other interior trim work is completed, the builder begins to install the tile.

Tile Installation

Builders install tile less frequently in new homes today than in the past. For example, many new homes no longer have ceramic tile baths. Instead, in place of ceramic tile, builders use one- or two-piece fiberglass tubs and shower enclosures along with vinyl floors.

Because ceramic tile is expensive to install, alternative materials and methods can reduce the cost of construction and thus go a long way toward making homes more affordable. However, builders do continue to use ceramic tile in many luxury homes for bath floors and walls, kitchen countertops, and even floor surfaces in some rooms. Other types of tile products, such as brick pavers or quarry tile, may also be used for decorative floor surfaces.

With the traditional approach, the builder installs ceramic tile over a continuous layer of mortar. As the first step in this approach, a worker trowels the mortar into reinforcing mesh attached to the wall or floor surface, much like plaster. The worker then applies the tiles to the surface and allows them to dry. (See photo 5.6.) The worker next returns and carefully

5.6 *Application of Ceramic Bath Tile*

grouts all the joints. This is both an expensive and time-consuming process.

Therefore, as an alternative, today most builders install tile to moisture-resistant gypsum board drywall with an adhesive, and the joints are grouted soon after. This approach eliminates much labor and drying time and results in lower installation costs.

The completion phase of constructing a house includes a few other tasks. At this point, for example, the builder installs the garage door, installs and trims the attic access door, and finishes the exterior painting.

At the end of the interior finish phase, the fireplace and other interior brickwork are in place. The builder has applied the plaster or drywall and installed the interior trim, including the door and window trim, moldings, cabinets, and finish hardware. After completing the construction of the interior staircases and applying the tile to appropriate places in the house, the builder can now turn to the final completion phase of house construction.

6

Completion
of the House

Everything may seem to occur at once during the last week or two of the homebuilding process. In fact, steps completed in this period of construction add about 25 percent to the value of the home. In addition, once the builder has completed the home and as soon as the lender approves the buyer's mortgage, everyone usually wants to go to closing as quickly as possible. For that reason, the builder will likely have more than one crew working at the same time during this period.

Some of the steps the builder may complete in the final days before closing include the installation of the wall and floor coverings, appliances, and light fixtures. At this point the builder also does the final installation of the plumbing and heating and cooling system before moving on to the final exterior improvements and landscaping.

Wall Coverings

The builder usually begins the final stages of house construction by installing the wall coverings, which include paint, stains, wallpaper, and vinyl wall coverings.

Paint and Stains. The builder generally tackles painting and staining first because no one wants paint or stains to get on the carpet! Most interior walls require a prime coat and a finish coat of paint. During this process, the painter retains some of each color to touch up the walls and the woodwork in case the paint is nicked during the final work on the house.

Generally, the builder has the walls painted with a flat latex, although the kitchen walls may be painted with a semigloss paint. These paints have a satin luster and are easier to clean. If you have not visited a paint store recently, you may find that it now offers a much wider variety of paints than in the past with some interesting choices in wall colors, textures, and surfaces.

Wallpaper and Vinyl Wall Coverings. Wallpaper and vinyl wall coverings can go up anytime after the painting and staining are finished. Before applying the wallpaper, the builder needs to prepare the wall with a special sealer. Without this sealer, the wallpaper will bond directly to the plaster or drywall. However, if the builder properly prepares the wall for the paper ahead of time, then years later when the homeowner wants to redecorate the wallpaper will be easier to remove.

Floor Coverings

Choice of floor coverings include hardwood, vinyl, and carpet. Of course, each category of floor covering has many varieties and subcategories. In fact, you may have visited a home that combines two or more floor materials in one room. Generally, hardwood and vinyl floorings go in first.

Hardwood Floors. A generation ago, builders used hardwood for all floors in homes. Unfortunately, this material and its installation have become relatively expensive over the years. However, use of tongue-and-groove hardwood flooring is still very popular in certain areas of the home like the foyer and family room.

The builder installs the hardwood before nailing the base molding in place, so the molding can cover the gaps between the hardwood and the wall. The strips of hardwood have a tongue at one edge and one end and a matching groove at the other edge and end. As the flooring is laid, the edges of the strips with the tongues fit into the edges of the strips with the grooves.

Installers usually lay down the hardwood flooring in stages. They nail the hardwood in place with a nail gun that drives special flooring nails through the tongues into the sub-flooring at an angle so that the nails are concealed when the next piece of flooring is installed.

Since hardwood is subject to variations in the humidity of the air, the builder must install this type of flooring under proper conditions. If the air is too humid when the hard-wood is installed, then it will shrink after installation and leave gaps. If the air is too dry, then the hardwood may expand and buckle later. The builder must carefully evaluate these factors when planning the installation of hardwood flooring. In fact, the builder may even want to heat the home in winter first to create the proper conditions for installing the hardwood floor.

After the hardwood floor is installed, the builder has it sanded and finished. The finisher sands the hardwood floor with a special floor sanding machine to the point where the floor is smooth and level. After sanding, the finisher thoroughly sweeps and vacuums the floor to remove the dust and particles. The finisher may then stain the floor and, after the stain has dried, apply one or more coats of clear finish to seal the floor.

In many homes today, builders use prefinished hardwood flooring. Since this type of flooring requires no sanding, staining, or finishing at the site, the builder can install it late in the construction process.

Vinyl Floors. After installing the hardwood floors, the builder turns to the process of laying the vinyl flooring. Vinyl roll or sheet goods have replaced linoleum as the material of choice in kitchens, utility rooms, playrooms, and other areas of heavy traffic in the house. This material requires a carefully prepared base or underlayment with a smooth surface.

Since vinyl usually comes in 6- or 12-foot-wide rolls, the vinyl installer joins the vinyl together with a seam between two sections. A good vinyl installer knows how to join the seams so the finished floor appears continuous. After installing a vinyl floor, the builder sometimes nails a small molding called a base shoe in place to cover the edges of the vinyl where the baseboard meets the floor.

Carpeting. Wall-to-wall carpeting goes in after all other interior finish work is completed. As the first step, the carpet installer tacks the padding to the subfloor. The carpet installer next lays sections of the carpet in place over the pad and pulls the sections tight to the edges of the room. The carpet installer then joins the sections of the carpet together with a special tape and uses an iron to melt the adhesive so the seams cannot be seen.

The builder will often direct the homebuyer to a full-line carpet supplier who can help the person select the right grade of carpet for each room, given the person's tastes, budget, lifestyle, and traffic patterns.

Appliances

After installing the wall and floor coverings, the builder can move on to installing the appliances in the house. These appliances may include a range, microwave oven, dishwasher (see photo 6.1), refrigerator, garbage disposal, washer, dryer, central vacuum, and trash compactor. The supplier usually delivers these appliances to the house.

You may remember that the builder has already installed the electrical service and plumbing lines for the appliances at the rough-in stage. As part of this process, the builder will have sized the kitchen cabinets to accept them. The only job left for the electrician and plumber is to hook up the wiring and plumbing and to ensure that the appliances operate properly.

If this sounds easy, remember that experience is an important factor here. Moving heavy appliances on a newly installed vinyl kitchen floor requires care, the appropriate moving equipment, and a strong back. In addition, although

6.1 Installation of Dishwasher

the refrigerator, stove, washer, and dryer may plug into receptacles, many built-in appliances don't necessarily just plug in. The electrician must then make any direct connections that are required.

Light Fixtures

Besides the appliances, at this point the builder has the light fixtures installed. Soon after signing the contract, homebuyers may go to a lighting showroom to select fixtures for the new home. Based on the house plans, the lighting showroom staff can help them choose fixtures that stay within their budget. Homebuyers need to do this as soon as possible after signing the contract, because if the showroom does not have certain light fixtures in stock, ordering and delivering them can take weeks or months.

In addition, selection of the light fixtures can affect the wiring. During the completion phase, the electrician picks the fixtures up from the showroom warehouse. After the builder has finished papering or painting the walls and ceilings, the

*6.2 Installation
of Light Fixture*

electrician then installs the light fixtures at the same time as the switches, receptacles, and doorbells. (See photo 6.2.)

Specialty subcontractors install many of the other electrical items, such as garage door openers or central vacuum cleaners. For those appliances the electrician usually only provides the needed power service. The electrician may also make the final connections to the heat pump or other types of heating and air conditioning systems. However, in some areas, depending on local customs, the electrician only runs the rough wiring and the heating contractor makes the electrical connection for these systems.

The electrician then notifies the municipal inspector that the home is ready for the final electrical inspection. In some jurisdictions, the official may perform this inspection separately from the final occupancy inspection. In either case, the power company must wait to turn on the power for the house until the local inspector has examined the electrical hookups and determined they are safe.

Final Plumbing Installation

About the same time that the light fixtures are installed, the builder begins the final plumbing installations. At this point the plumber has to complete three tasks. First, the plumber needs to connect appliances like the dishwasher or washer.

Second, the plumber must install the plumbing fixtures and faucets. Like the light fixtures, the homebuyer must work with the builder to select the plumbing fixtures early in the process. Installing the plumbing fixtures includes setting the toilets, sinks, and other fixtures; hooking up the drains and water lines; and testing to make sure the system has no leaks.

The builder often installs toilets, bathroom vanity cabinets, and kitchen cabinets after installing vinyl floor covering. Ceramic or quarry tile floors, on the other hand, are usually installed after toilets, cabinets, and fixtures.

As the third and final step, the plumber also needs to complete the hookups to the public sewer and water service or to the well and septic tank.

Sewer and Water Service

The plumber's responsibility regarding the sewer and water hookups varies by community. In many locations, the local sewer and water authorities provide a tap or connection into the sewer or water main.

Before the first home is built in a planned subdivision, the developer often installs a sewer lateral or branch to each lot. The sewer lateral is a 3- or 4-inch underground pipe, which is usually made of heavy plastic or cast iron and left capped until the builder constructs a home on the lot. At the appropriate time, the plumber digs a trench from the house to the end of the sewer lateral and ties the house into the community sewer.

Sewage generally needs to run downhill, although neighborhoods in hilly areas can use pumping stations to compensate for the lack of gravity-assisted flow. In addition, wherever possible, the sewage service from individual homes should run downhill. However, if a home is located downhill from the sewer, then it may also need some type of sewage pumping system. This is usually a self-contained, collection tank with a built-in electric pump that acts like a sump pump and activates when the tank fills up. These tanks are usually located in a pit in the front or rear yard.

Unlike the sewer service, the water service does not depend on the flow of gravity. A common location for the water main is under the center of the main street or within the right-of-way of the main street on one side or the other.

When a subdivision is proposed, the developer usually agrees to provide water service to each lot, generally through a ¾- or one-inch diameter plastic or copper pipe that runs from the water main to the front yard of the proposed lot. At the appropriate time, the plumber then runs the pipe to the house. Later the water authority sets the meter, usually after the homeowner pays a deposit.

Septic Tanks and Wells

In some communities public sewer and water service are unavailable. The process for providing private sewer and water service varies by locale. Generally, a permit must be obtained from the local health department. The builder then hires specialty subcontractors to drill the well and install the septic tank, which is an onsite method of household waste disposal.

In the case of a septic tank, two things are actually needed. First, the septic tank itself is a large, concrete or fiberglass tank, located underground near the house, that collects the sewage. Bacteria in the septic tank turn the outflow into a liquid.

The second thing required for a septic tank is a leach or drain field. The drain field is a network of trenches, filled with gravel and perforated pipes, that allows the liquid waste to drain into the soil. (See drawing 6.3.) A properly operating septic system is odorless and efficient and will last for years with proper maintenance.

Before the builder installs a septic system, however, the lot must pass a percolation test administered or regulated by the local health department. With this test, the inspector care-

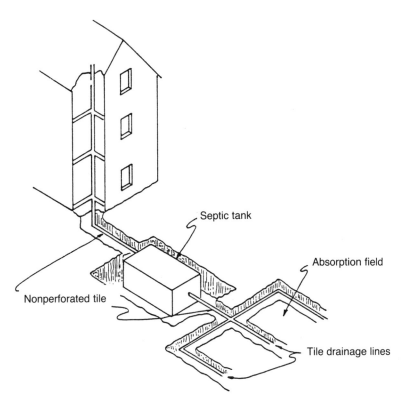

Septic tank

Absorption field

Nonperforated tile

Tile drainage lines

6.3 Typical Septic System

fully measures the rate at which the soil absorbs the water on the lot. If the soil absorbs the water fast enough, then the inspector is confident that the septic field will work and approves the installation of the system. As part of the testing process, the inspector may recommend a specific design for the drain field or may approve or modify a design laid out by the builder or designer.

If a site needs a well, a well driller can dig a well on the property and provide a pump to bring the water to the surface. However, the driller cannot guarantee that water will be found or, once found, that it will be drinkable. For that reason, before having a well dug, the prospective homeowner should check with other homeowners in the neighborhood about their experiences with the water supply. Generally, all of the wells in a given neighborhood tap into the same aquifer or underground source of water. However, depth, volume, and water quality can still vary among neighboring wells.

The local health department may also require that an independent laboratory test a water sample from the well before it is certified. Once the well is certified, the homeowner may still want to test the water periodically. If the water is too hard or contains too many minerals, then a relatively inexpensive water softener system may cure the problem.

Besides the final electrical and plumbing work, as part of the completion of the house the builder also finishes the final hookup of the heating and cooling system.

Final Hookup of the Heating and Cooling System

Chapter 4 described several options for heating and cooling the home. The type of system was selected early, since the builder had to install many of the components at the rough-in stage. As you may remember, the two major choices for heating and cooling the home were an electric heat pump or a gas furnace with an electric air conditioner. At this point, the builder completes the final hookup of those systems.

Electric Heat Pump

If an electric heat pump is the primary source of heating and cooling for the house, then the builder installed several items at the rough-in stage, including the following:

- ductwork, usually either under the home or in the attic
- standard 120- and 240-volt wiring to the planned location of the air handler
- 240-volt wiring to the location of the heat pump itself, typically in an unobtrusive place outside the home
- low-voltage thermostat wiring between the planned location of the air handler and the thermostat

The final installation of an electric heat pump involves two steps. The first step is to set the air handler in the house. This is a box that primarily functions as an air handler that houses the heat pump coil and auxiliary electric heating elements. The heat pump heating capacity is supplemented with electric heating coils that function when the weather is too cold for the heat pump to operate effectively.

As part of the first step, the builder also connects the air handler to the two types of ducts running under or in the house—the supply ducts and the return ducts. The supply ducts distribute heated or cooled air throughout the home, while the return ducts bring air back to the air handler for reconditioning.

As the second major step to installing a heat pump, the builder sets the heat pump unit outdoors on a mounting pad or platform. The builder then connects the refrigerant lines between the outdoor heat pump compressor and the refrigerant coil on the air handler. (See photo 6.4.) Heat pumps that combine all functions in a single unit are also available for mounting on a pad or platform outside the house or on the rooftop.

6.4 Installation of Electric Heat Pump Compressor

Gas Furnace and Electric Air Conditioner

The final installation of a gas furnace with an electric air conditioner also has two basic steps. If the house has a typical gas furnace and an electric air conditioner combined into one system, the first major step is to install the furnace in the house, mount the cooling coil in the supply duct directly above the furnace, and connect the furnace to the duct system, much like the air handler used with a heat pump. As a second step, the builder then sets the outdoor compressor unit, again much like a heat pump except that this unit only functions for cooling. The refrigerant lines are then connected between the outdoor unit and the cooling coil above the furnace.

Other types of gas furnaces are available, which the builder can mount in an attic or crawl space or on the roof or a platform in the backyard. These units can also be used in combination with electric air conditioning.

The builder now has a certified gas plumber connect the gas supply line from the gas meter to the gas furnace. If the furnace is the only gas appliance in the home, then the gas line will run from the meter directly to the furnace. If the home has other gas appliances, such as a gas range or water heater, then the supply lines may branch out and run through the crawl space, basement, or walls to the various appliances. Once the plumber has installed the gas lines, the builder has the local authority inspect them.

Although central heating systems in the form of an electric heat pump or a gas furnace are the most common, other systems are also available. Furnaces may be fired with natural gas, liquefied petroleum gas, or fuel oil. Baseboard electric heaters are sometimes used because they have a relatively low cost for initial installation. Particularly in the Northeast, hot water baseboard (hydronic) heaters connected to a gas or oil-fired boiler are also popular. The builder can advise the homebuyer on what kind of heating system might provide the most comfort at the lowest installation and operating costs for a given climate.

At this point, house construction is essentially complete. The delivery trucks have come and gone and the dump trucks have hauled the trash away. Now the exterior of the house receives its finishing touches, including the installation of decks, patios, driveways, and walkways. At the very end of the process, the lot is landscaped. This is a general term for the final grading of the site, the laying of sod, and the planting of trees, shrubs, and grass.

Final Exterior Improvements

Before landscaping, the builder completes several exterior improvements to the house. Besides the construction of decks and patios, the driveway, and walkways, these improvements can also include the construction of screened porches, stoops, steps, and other attachments that may enhance the entry to and exit from the house.

Decks and Patios

Decks and patios are options that extend a home's living space. The configuration of the house and yard, budget, and personal preference of the homebuyer determine the size and shape of decks and patios. Zoning laws, local ordinances, subdivision covenants, property owner association rules, and local building codes may also affect their location, size, shape, or design. In addition, climate may affect the construction of patios and decks. For example, decks or patios with light-weight or low, sloping roofs may not be appropriate in windy or snowy climates.

Decks are elevated outdoor platforms, usually constructed of pressure-treated or durable wood, such as cedar or redwood, and edged by a railing for safety and convenience. (For a picture of a recently completed deck, see photo 6.5.) On the other hand, unlike decks, patios are built directly on the ground.

Common materials for patios include concrete, brick, or stone. A variety of textures in each of these materials can give dramatic style and design to an otherwise plain patio. As the first step in constructing the patio, the builder creates forms out of 2x4s or 1x4s. The surface of the patio must

6.5 Completed Deck

have a gradual slope away from the house to assure proper drainage.

Next the builder has the concrete crew pour the slab, usually 4 inches thick. The builder may include reinforcing wire mesh in the slab to minimize cracking. The patio should not be poured in very cold weather because it will not cure properly. On the other hand, concrete poured in very hot weather tends to set before the crew can properly work and finish it.

Driveways

Depending on the size and shape of the house and lot, driveways can be long, short, straight, curved, or circular. The builder often has the driveway constructed of concrete in a manner similar to the patio. However, since the driveway will carry greater loads than the patio, the builder may place a firm base material—usually sand, gravel, or crushed stone—underneath the layer of concrete. As an alternative, in areas where soil conditions are very stable, the concrete driveway may be placed directly on the soil. (See photo 6.6.)

6.6 Driveway under Construction

Other common driveway materials include the following:

- asphalt, which the builder places over a compacted layer of aggregate
- brick pavers, which the builder usually installs over a sand or rough concrete bed
- loose stone

When loose material is selected for the driveway, the builder may set in place a type of edging, such as pressure-treated 1x3s or 1x4s, to keep the material within the boundary of the driveway.

Walkways

The house plans may also call for one or more walkways. Possible materials for the walkway include concrete, brick, flagstone, and asphalt. In established neighborhoods with sidewalks, the walkway may lead from the front door through the front yard to the sidewalk.

In other homes, the walkway may lead from midway up the driveway through the front yard to the front door. A builder may also build a walkway from the driveway around the side of the house to a side or rear door or patio. This type of walkway can help to direct traffic away from the front door and more formal areas of the house.

Landscaping

Once the builder has laid the patio, driveway, and walkway, the finishing touches are put on the landscaping. As the first step in landscaping, the builder uses a bulldozer or similar machine to grade and smooth the lot. As part of this process, the builder redistributes the earth left from excavating the basement. The earth should slope away from the house to assure effective drainage. Later the homeowner must care for this slope to maintain proper drainage.

In general, landscaping requires specialized skills and knowledge. In some areas, some builders may prefer that the homebuyer contract separately with a landscaping specialist to complete this final stage. In other areas, the municipality may require the builder to sow grass seed or sod the yard to assure against erosion.

Because of the appeal of trees to homebuyers, some builders may plant trees as part of their final landscaping efforts. Trees can provide decoration, privacy, shade in the summer, and protection against wind in the winter. They can also help to buffer noise and control erosion. Finally, planting trees may enhance the value of a new home if included in the final landscaping.

After the builder has completed the house and before the new homebuyer can occupy the house, the local building department must perform a final inspection of the plumbing, heating and air conditioning, and electrical systems, and the interior and exterior of the house. Once the building inspector determines that the house has been built in compliance with all applicable code and zoning requirements, a certificate of occupancy is issued and the house is ready for the builder's and homebuyer's final inspection.

Conclusion

Typically, just before or on the day of closing, the homebuyer and the builder walk through the house to make sure everything is satisfactory. During this walk-through, the builder and homebuyer check for such things as whether the paint needs some final touch-ups, the carpet has a loose seam that needs tacking down, or a cabinet drawer sticks. The buyer and builder may spot several such minor items that will not prevent closing but need to be agreed upon. The builder may have a worker on hand with the necessary tools and materials to fix minor items on the spot or the builder and homebuyer may simply agree on the items to be fixed and when they will be completed.

As the homebuyer and builder walk through the house, most builders ask the buyer to make and sign a list—called a punchlist—that identifies anything the builder needs to address after closing. The builder may also ask the homebuyer to keep a second list of other items after the move-in. When a reasonable time has passed for the new homeowner to spot such items, the builder usually sends a worker to take care of them.

Even the best builders may need to return and fix a few minor items in a new home, and the new homeowner should expect this as a normal part of the process. If the home builder is backed by a recognized warranty program, then that program usually provides in general terms that the builder will resolve reasonable problems that may arise in the workmanship or materials of a new home over a specified period—perhaps a year. This is considered a reasonable time for the average homeowner to spot and report such repairs. In addition, some warranty programs may cover certain systems in the house, like the plumbing, for a longer period. Finally, major structural components of the house, like the foundation or roof, may be warranted for a more extended period.

Beyond the warranty coverage, however, the homeowner also has a very important responsibility to maintain the new home in proper working order. Proper maintenance of a home will minimize the major repairs over time and increase the homeowner's enjoyment of the home.

So, with the help of many skilled craftspeople, the builder has pieced together a complex puzzle of elements that go to make up the completed home. As part of this process, the builder has prepared the site; laid the foundation; framed the structure of the house; installed the plumbing, heating, and electrical systems; completed the interior and exterior finish; and landscaped the site. Once the certificate of occupancy is issued and the builder and homebuyer have successfully completed their walk-through, the home is ready for closing.

Buying a new home is the largest and possibly best investment most Americans will ever make. It is hard to find a financial adviser who does not recommend homeownership as the first step in planning for a family's financial security. Buying a newly built home also gives the homebuyer the latest in materials, appliances, equipment, design, and amenities available in homebuilding today and helps to assure the best possible resale price when it's time for the homeowner to make that move-up buy.

Glossary

asphalt A dark, tarlike bituminous material commonly used in house construction for roofing, waterproofing, dampproofing, exterior wall covering, and pavement.

backfill Earth or other material used to fill in around foundation walls, usually built up to drain water away from the foundation.

base molding A decorative band of finish board used to cover the joint between the wall and floor. Also called baseboard or base.

building code Minimum legal requirements for all aspects of construction, established and enforced by local governments to protect public health and safety. Building officials and others with firsthand knowledge of construction practices establish these codes.

cast-in-place A term used to describe concrete that is poured between wooden or metal forms to harden in place.

cellulose Recycled wood fiber used in house construction for insulation and flooring material. Can be chemically treated with flame retardant.

center beam A wood or steel member that runs the length of the first floor of a house and supports the house structure above it. The center beam bears on the foundation wall at each end of the house and is supported along its length by columns or piers.

certificate of occupancy   A legal document, issued by a building inspector, stating that a house has passed all inspections and is ready for occupancy.

chair rail   A band of molding applied at chair-back height along a wall to protect the wall finish from chairs pushing up against it. Also used as a decorative detail.

circuit breaker   A safety feature for each electrical circuit in the distribution panel. Should the demand for electricity on a particular circuit become excessive, the circuit breaker automatically cuts the flow of electricity through that circuit. The electrical circuit remains broken until the circuit breaker is reset. Circuit breakers serve the same purpose as fuses in older electrical circuits.

collar beam   In roof framing, a horizontal piece that provides structural strength by connecting opposite rafters.

concrete   A mixture of Portland cement, sand, gravel, and water that hardens into a rocklike mass.

concrete block   Precast hollow or solid building block made of cement, water, and aggregate, such as sand or gravel. Commonly used in wall construction.

cornice   On the exterior of a house, structural trim where the roof and walls meet. Also called soffit when the trim overhangs the walls.

crawl space   In houses without basements, the space between the ground surface and the first floor sufficient to crawl around in for utility installation and repairs.

crown molding   A decorative molding used to cover the joint between the wall and ceiling.

distribution panel   A metal box through which all electrical wiring passes, usually located at the juncture with the utility company line. The panel distributes electricity to all usage points in a house by means of circuits or sets of wires. This is where the circuit breakers are located. Also called a breaker box.

dormer   A projection built out from a sloping roof as a room extension or for a window.

drainage pipe   Plastic pipe or clay tile, sometimes perforated, laid in the ground to carry water away from areas around the foundation. Also called drain tile.

drywall   Paper-faced gypsum board panels used in interior wall finishing instead of plaster. Also called gypsum board or gypsum wallboard.

duct, ductwork   Round or rectangular conduits usually formed of sheet metal and used to transport heated and cooled air from heating and cooling equipment to the various rooms in a house.

eaves   The edge of a roof that runs parallel with the ground.

excavation   Removal of earth or rock, as for the basement of a house.

fiberglass   A nonflammable material made of glass fibers and used in thick woollike blankets as building insulation and as a reinforcement in some plastic fixtures.

flashing   Sheet metal or plastic used to cover joints and openings in exterior surfaces of the house to protect against water leakage.

footing   A base, usually made of concrete, beneath foundation walls, columns, piers, and chimneys. Designed to distribute the weight of these elements to the soil, depending on the bearing capacity of the soil.

formwork   Support structure for freshly poured or cast-in-place concrete.

foundation   Part of the house that transfers the load of the house to the ground, usually made of concrete.

framing   The process of constructing the structural skeleton of a house, usually made of wood or steel studs, beams, and joists.

gable   The triangular end wall of a house that extends from the eaves to the peak of the roof.

grading   The preparation of a site by cutting, filling, or both to accommodate construction of a house. Also, filling in with earth or other material around a completed house at a slope to direct water away from the house.

gutter   Metal, plastic, or wood channel at the eaves of a house, usually sloped slightly to carry off rainwater and snow melt.

gypsum board   Panels used in interior wall finishing, consisting of mineral gypsum pressed between two layers of heavy paper. Also called drywall and gypsum wallboard.

heat pump   Combination heating and cooling equipment. In winter, the heat pump extracts heat from outdoor air, which is circulated through the house. In summer, the heat pump extracts heat from indoor air, which is discharged outdoors. The cooled air is then circulated through the house.

HVAC   Common building industry abbreviation for heating, ventilation, and air conditioning systems.

inspection   Examination of work completed on a structure to determine compliance with building code and other requirements.

insulation, electrical   Any material used to cover electrical conductors.

insulation, thermal   Materials used in house construction to retard heat flow or protect against sound transmission or fire.

joint compound   A pastelike material used to cover fasteners and joints in drywall for a smooth finish. Also called joint cement.

joists   A series of horizontal parallel beams that support floors and ceilings.

load bearing   Providing support for the weight of a house or other loads such as people, furniture, and snow.

masonry   General construction term for materials set in mortar, including stone, brick, concrete, tile, and glass block.

mechanical systems   General term for heating and air conditioning. Sometimes also applied to include plumbing and electrical systems.

molding   Wood, metal, plastic, or plaster trim used around windows and doors, at the tops and bases of walls, along cornices, and for other decorative details.

mortar   A thick, pastelike material that hardens to bond masonry units together. Usually made of a mixture of Portland cement, sand, lime, and water.

oriented strand board   A structural panel composed of layers of sliver-like wood strands bonded together with resin. Each layer consists of compressed strands oriented at 90 degrees to the previous layer.

percolation test   A soil test used to determine the rate at which water is absorbed into the ground. Results are used to establish the best locations and required size for a septic field on a piece of property. Also called a perc test.

permit   A document issued by a local government agency allowing construction work to be performed in conformance with local codes. Work may not begin until the builder has obtained the required permits, and each permit-issuing agency must inspect the work at certain specified points during construction.

plywood   A type of building material made by gluing three or more thin layers or plies of wood together in panels. Plies are laid so that the wood grain alternates 90 degrees with each successive layer. This increases the plywood's overall strength and counteracts shrinking, swelling, and warping in each ply.

polyethylene   A durable, pliable, waterproof plastic film used as a moisture vapor barrier in house construction.

polyvinyl chloride (PVC)   Rigid, durable plastic material used in plumbing for pipes and fittings.

pressure treated   A process of forcing chemicals into wood to prevent deterioration from rot, mold, termites, and other wood-destroying insects.

prints   Complete construction plans, drawn to scale, used by builders and subcontractors to build a house. Prints usually include the site plan, foundation plan and cross section, floor plans, elevations, building and wall cross sections, mechanical systems, and special construction details. Formerly called blueprints.

R-value   A term that, when used with a number, indicates the level of resistance to heat flow in a building material. The higher a material's R-value, the more effective insulation it provides.

rafters   The structural members that form the legs of the triangle created in roof framing. Rafters are joined at the peak of the triangle by the ridge board and support roof sheathing and roofing materials.

ridge board   The length of lumber at the peak of a roof to which the upper ends of the rafters are fastened.

rough-in   The stage of house construction that follows framing, when the builder installs all systems that will be concealed behind the walls, including plumbing, heating and air conditioning ducts, and electrical wiring.

septic system   A sewage disposal system for individual homes. A holding tank for raw sewage is installed in the ground, where bacteria break down and liquefy the sewage. The liquid waste is then discharged to a drain field where it slowly passes into the soil and is eventually purified.

setback   The minimum allowable distance between a structure and its lot lines established by local zoning ordinances.

sewer   A system of pipes carrying away storm runoff or carrying sewage to a municipal processing plant.

shake   Hand-split wood shingle.

sheathing   Sheets of plywood, flakeboard, oriented strand board, insulation board, or other materials used to cover the exterior of a house's frame. A finish siding material is then applied over the sheathing.

shingles Roof or wall covering of asphalt, wood, tile, slate, or other material applied so as to shed rainwater.

siding The exterior finish of a house applied over the sheathing and generally made of wood, vinyl, steel, aluminum, or other durable, decorative materials.

sill A wood support, usually a 2x6, laid flat on top of the foundation wall and used as a base for floor framing; also called the sill plate. Also, the member forming the lower side of an opening, such as a windowsill or doorsill.

slab A flat layer of poured concrete.

soffit Exposed underside of a projecting building part such as a roof overhang.

stack vent Vertical pipe in a plumbing drainage system, which extends up through the roof to relieve pressure differences that could otherwise siphon water out of plumbing traps in the house.

stake-out Measuring house dimensions on a lot in accordance with the house plans and using stakes to indicate each corner.

stucco A cement plaster used as an exterior wall finish.

stud An upright wood or metal member used to frame walls and partitions.

subcontractor A person or company that contracts with the builder to perform work on a specific part of a construction job, such as excavation, plumbing, electrical work, or landscaping. Also called sub.

subflooring Rough boards, plywood, waferboard, or oriented strand board installed on top of the floor joists, over which the finish floor is laid.

sump pump Device used to remove liquid from a drainage pit or sump.

termite A wood-devouring insect that can demolish the woodwork of a structure.

transformer A device for increasing or decreasing electrical voltage. Used by electrical utilities to convert high voltage levels to 120/240-volt service for houses.

trap In plumbing, a bend in a waste pipe designed to hold water. The water acts as a seal to prevent insects and harmful sewage gases from gaining access into a house.

trench A narrow excavation in the earth for the installation of footings, pipes, drains, and electrical cables.

trim Finish details and materials, particularly moldings around windows and doors; moldings at joints, intersections, and corners; and other decorative work on the house.

truss Preassembled roof framing member, made of wood and commonly manufactured in a triangular shape, that replaces ceiling joists, rafters, and collar beams.

underlayment Moisture-resistant material, such as an asphalt-treated paper, applied over roof and wall sheathing under roof and exterior finish. Also a material used over subflooring to provide a smooth surface for a finish flooring.

utilities Services available to citizens of a community, such as water, electricity, gas, and sewage disposal.

vanity A dressing table; also, a wash basin with an enclosed cabinet below.

vapor barrier Treated paper or plastic film that retards the flow of water vapor.

veneer Any decorative, nonstructural surface layer.

wallboard See gypsum board or drywall.

zoning Division of a county or municipality into land use categories. Establishment of regulations governing the use, placement, spacing, and size of land parcels and buildings in each category.

Additional Reading

Advanced Framing: Techniques, Troubleshooting & Structural Design. Richmond, VT: The Journal of Light Construction, 1992.

Basic Plumbing by Sunset Books. Menlo Park, CA: Lane Publishing Co., 1992.

Building Your Own Home by Wasfi Youssef. New York: John Wiley & Sons, 1988.

Dreams to Beams: A Guide to Building the Home You've Always Wanted by Jane Moss Snow. Washington, DC: Home Builder Press, National Association of Home Builders, 1988.*

Estimating for Home Builders by Jerry Householder and John C. Mouton. Washington, DC: Home Builder Press, National Association of Home Builders, 1992.*

Fine Homebuilding Questions and Answers about Building. Magnolia, MA: Peter Smith Publishing, Inc., 1991.

Foundations and Masonry by Fine Homebuilding. Newton, CT: The Taunton Press, Inc., 1990.

Frame Carpentry by Fine Homebuilding. Newton, CT: The Taunton Press, Inc., 1990.

Fundamentals of the Construction Process by K.K. Bentil. Kingston, ME: R.S. Means, 1989.

Houses: The Illustrated Guide to Construction, Design & Systems by Henry S. Harrison. Chicago: Real Estate Education Company and Residential Sales Council™ of the Realtors National Marketing Institute,® 1992.

How to Design and Build Your Own House by Lupe DiDonno and Phyllis Sperling. New York: Alfred A. Knopf, Inc., 1987.

Land Development. Washington, DC: Home Builder Press, National Association of Home Builders, 1987.*

Roof Framing by Marshall Gross. Carlsbad, CA: Craftsman Book Company, 1984.

The Complete Guide to the Home Remodeling and Construction Process by J. Hardy LeGwin. West Newton, ME: J.H. LeGwin Associates, 1990.

Understanding Building Codes and Standards in the United States. Washington, DC: Home Builder Press, National Association of Home Builders, 1989.*

Using Building Systems: Modular, Panelized, Log, Dome by James Carper with the NAHB Building Systems Council. Washington, DC: Home Builder Press, National Association of Home Builders, 1989.*

What Every Potential Homeowner Should Know about Construction, vol. 1, by Douglas E. Hedlund. Tucson, AZ: Condata Co., 1989.

Wood Frame House Construction. Washington, DC: Home Builder Press, National Association of Home Builders, 1988.*

---

*For more information about books published by Home Builder Press of the National Association of Home Builders, call (800) 223-2665 or write Home Builder Press, National Association of Home Builders, 1201 15th Street, NW, Washington, DC 20005-2800.